Abnormal States of Brain and Mind

Readings from the *Encyclopedia of Neuroscience*

Abnormal States of Brain and Mind
Selected and with an Introduction by J. Allan Hobson

Comparative Neuroscience and Neurobiology
Selected and with an Introduction by Louis N. Irwin

Learning and Memory
Selected and with an Introduction by Richard F. Thompson

Sensory Systems I: Vision and Visual Systems
Selected and with an Introduction by Richard Held

Sensory Systems II: Senses Other than Vision
Selected and with an Introduction by Jeremy Wolfe

Speech and Language
Selected and with an Introduction by Doreen Kimura

States of Brain and Mind
Selected and with an Introduction by J. Allan Hobson

Readings from the
Encyclopedia of Neuroscience

Abnormal States of Brain and Mind

Selected and with an Introduction by
J. Allan Hobson

A *Pro Scientia Viva* Title

Birkhäuser
Boston · Basel

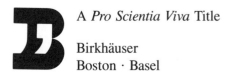

Library of Congress Cataloging-in-Publication Data
Abnormal states of brain and mind.
(Readings from the Encyclopedia of neuroscience)
Bibliography: p.
1. Neurology. 2. Psychiatry. 3. Neuropsychiatry. I. Hobson, J. Allan, 1933–
RC 341.A26 1989 616.8 88-7822
ISBN 0-8176-3397-9

This collection is made up of articles from the *Encyclopedia of Neuroscience*, edited by George Adelman.

Printed and bound by Edwards Brothers Incorporated, Ann Arbor, Michigan.
Printed in the United States of America.

9 8 7 6 5 4 3 2 1

ISBN 0-8176-3397-9
ISBN 3-7643-3397-9

Series Preface

This series of books, "Readings from the *Encyclopedia of Neuroscience*," consists of collections of subject-clustered articles taken from the *Encyclopedia of Neuroscience*.

The *Encyclopedia of Neuroscience* is a reference source and compendium of more than 700 articles written by world authorities and covering all of neuroscience. We define neuroscience broadly as including all those fields that have as a primary goal the understanding of how the brain and nervous system work to mediate/control behavior, including the mental behavior of humans.

Those interested in specific aspects of the neurosciences, particular subject areas or specialties, can of course browse through the alphabetically arranged articles of the *Encyclopedia* or use its index to find the topics they wish to read. However, for those readers—students, specialists, or others—who will find it useful to have collections of subject-clustered articles from the *Encyclopedia*, we issue this series of "Readings" in paperback.

Students in neuroscience, psychology, medicine, biology, the mental health professions, and other disciplines will find that these collections provide concise summaries of cutting-edge research in rapidly advancing fields. The nonspecialist reader will find them useful summary statements of important neuroscience areas. Each collection was compiled, and includes an introductory essay, by an authority in that field.

George Adelman
Editor,
Encyclopedia of Neuroscience

Contents

Introduction

The subjects of brain and mind continue to fascinate philosophers. Are they really two entities? If so, how do we conceive of their relationship? If not, why do we persist in using different names for them? Should we not begin, at least tentatively, to refer to a brain–mind (or a mind–brain)? This "mind–brain problem" continues to divide clinicians. Psychiatrists treat what they call mental illness, as if such states were not also disorders of the brain. Neurologists diagnose brain diseases without considering subjective experience as significant manifestations of brain function. How did this mind–body schizophrenia arise? And can it be cured – or at least treated?

In the nineteenth century psychiatry and neurology were but one branch of medical science, and it was called neurology. Hughlings Jackson, a distinguished neurologist, was superintendent of the West Riding Lunatic Asylum in Yorkshire, where his observations of patients led to his dynamic theories of the evolution and dissolution of brain and behavior. And when William James published his *Principles of Psychology* in 1890 he began with what was then known about the brain. Sigmund Freud, the founder of psychoanalysis and the proponent of the psychodynamic approach to mental illness, was trained as a neurologist and worked in that capacity until his clinical interest in hyponosis and hysteria convinced him of the need to develop a psychology that was independent of neurology. Freud concluded that he could not build a psychological theory upon what was known about neurons in the 1890s and proceeded to elaborate his psychoanalytic theory as if it had no relationship to brain science. The resulting separation of neurology and psychiatry reflected a paradoxical and unproductive resurgence of brain–mind dualism that has plagued medicine through the first half of the twentieth century.

With the recent explosive growth of neurobiological science, we are now witnessing renewed interest in brain–mind integration. Psychiatric research has come to look more and more neurologic. And with the growth of behavioral and cognitive science, neurology has a renewed interest in many of the same clinical conditions that it earlier relegated to psychiatry. Hughlings Jackson, William James, and even the youthful Freud, would all be pleased by this recent turn of events. Much of the groundwork for this clinical revolution is to be found in the basic science topics covered in a companion set of selections from the *Encyclopedia of Neuroscience* called *States of Brain and Mind*.

This clinical volume abounds with encouraging examples of cross-field fertilization. The 50 selections cover the major categories of clinical dysfunction and the treatments used to deal with them. Among the dysfunctions, a distinction may be usefully made between "diseases," – where the pathogenesis is known to be structural (and often genetic) – and "disorders," where the pathogenesis is either unknown or known to be functional, hence situational, or environmentally determined. By this definition there are only five diseases:

Alzheimer's Disease
Creutzfeldt-Jakob Disease
Down Syndrome
Huntington's Disease (HD)
Phenylalanine and Mental Retardation (PKU)

Each of these entities is characterized by mental defects that result from a well-established neuropathology.

Many of the conditions covered may well be diseases, but in most cases we don't yet know enough to say. In this category we consider the following entries:

Autism
Dyslexia
Epilepsy
Gilles de la Tourette Syndrome
Mental Illness, Genetics of
Mental Illness, Nutrition and
Mental Retardation
Mood Disorders
Schizophrenia

This list contains most of the problems at the cutting edge of biological research and the patients affected by these disorders, not surprisingly, are of interest to both neurologically oriented psychiatrists and behavioral neurologists.

Other topics included are either the manifestations of well-known neuropathologies or may be the nonspecific outcomes of a variety of functional causes. In this group of "symptoms and signs" we consider:

Amnesia
Coma
Deafness
Dementia
Eating Disorders
Eye Movement Dysfunctions and Mental Illness
Psychosis
Sleep Disorders

Some of the items listed here bridge the gap between mind and body. For example, amnesia and psychosis may both be entirely functional, as well as sometimes symptoms of organic brain disease.

The following list of states is even more clearly functional and includes conditions that are even sometimes entirely normal:

Addiction
Aging of the Brain
Alcoholism
Anxiety and Anxiety Disorders
Fetal Alcohol Syndrome
Premenstrual Syndrome
Stress, Neurochemistry of
Substance Abuse
Tolerance and Physical Dependence

The heterogeneity of all these topics and the blurred boundaries between them helps us understand why no rigid categorical system of classification can work. And it may help explain my emphasis on the concept of state in this volume's title. It seems quite likely that *all* of the above conditions are altered states of the brain–mind. What we need, then, is a more clearly formulated state concept, not a more loosely constructed concept of disease.

Some of the entries in this volume describe whole fields that turn out to be highly treatment-oriented. The wide range of treatments, from psychoanalysis (the talking cure) to psychosurgery (lobotomy, lobectomy, cingulotomy, etc.)—both of which are often recommended for the same disorders!—is a sign of our still monumental ignorance and the obdurate difficulty of the subject matter:

Behavior Therapy, Applied Behavior Analysis, and Behavior Modification
Behavioral Medicine
Convulsive Therapy
Neuropharmacology
Psychiatry, Biological
Psychoanalysis
Psychopharmacology
Psychosurgery

The last—and longest—category is a pharmacopeia of the chemicals that change our brain-minds. Some are prescribed by doctors and some are sold on the streets for recreational purposes. In either case their use is almost always problematic because of side effects, addiction and toxicity:

Amphetamines
Antidepressants
Cocaine
Hallucinogenic Drugs
Heroin (Diacetylmorphine)
Lithium in Psychiatric Therapy
Marijuana
Monoamine Oxidase (MAO) Inhibitors in Psychiatric Therapy
Morphine
Neuroleptic Drugs
Phencyclidine

The net result is a collection of papers that is more a scientific casebook than a comprehensive textbook of psychiatry-neurology. Being research-oriented, this compendium should be useful as an adjunct to the texts of both fields, and it may thus serve well as a reader to supplement medical school or undergraduate courses. We construe it also as a potentially valuable guide to laymen seeking recent scientific views of all-too-common clinical problems.

J. Allan Hobson, M.D.
Professor of Psychiatry
Harvard Medical School

Addiction

Harold Kalant

Addiction (or drug dependence, as the World Health Organization now recommends that it be called) is a concept that was originally clear in empirical terms, but became progressively more confused by successive attempts at official definitions. In North America there is a tendency to regard physical dependence, as revealed by a withdrawal reaction, as the cardinal feature of addiction, but this is putting the cart before the horse. The essence of addiction is drug-seeking and drug-taking behavior that has become a central element of the individual's life pattern. If the frequency and amount of drug taking are sufficiently high, and the circumstances of use are appropriate, tolerance and physical dependence are likely to result, and social, psychiatric and medical problems of various kinds may be produced, but these are all consequences of addiction. The fundamental question is: What causes the drug-taking behavior to become so strongly established as to generate these consequences?

Dependence and reinforced behavior

Currently, the most widely accepted behavioral analysis of addiction is based on the tenets of operant behavioral psychology. Self-administration of a drug gives rise to two types of effect that are held to be essential for producing dependence: (1) discriminative stimuli, i.e., those distinctive internal cues by which the drug taker can recognize the drug that has been taken, and (2) reinforcing effects (in nonoperant terms, rewarding effects), which increase the probability of repeating the self-administration. Most drugs, even highly reinforcing ones, also have aversive (i.e., punishing) effects which tend to reduce the probability of repeated self-administration. Presumably the net balance between reinforcing and aversive effects determines the dependence liability. The distinction may not be absolute: some investigators postulate that effects which are usually aversive may, in some circumstances, be reinforcing to some users.

Methods of study. The discriminative properties of a drug can be compared experimentally with those of other drugs, in order to determine the degree of similarity or difference. In humans this can be done in double-blind trials by giving the subjects small doses of standard drugs and placebo, and asking which one the test drug most closely resembles in its subjective effects. Experimental animals are trained to obtain food rewards by pressing the correct lever in a two-lever operant chamber. One lever is correct when the animal is performing under the influence of a small dose of a standard drug, and the second lever is correct when the animal has received a different standard drug or placebo. The animal must then perform under the influence of a test drug, and the percentage of total responses it makes on each of the two levers gives a measure of the degree to which the animal perceives the effects of the test drug as similar to, or different from, those of the standard drug(s) or placebo.

The reinforcing properties of drugs can also be studied experimentally. The direct method consists of fitting the animal with an indwelling venous catheter connected to a reservoir of drug solution by an infusion pump which is activated when the animal presses a lever in the operant chamber. By appropriate choice of drug concentration, and programming of the pump, the experimenter can cause a preselected dose of drug to be injected in response to any desired pattern of bar pressing by the subject (i.e., the experimenter can set up a schedule of reinforcement). The more work an animal will expend to obtain a dose of drug, the more strongly reinforcing the drug is inferred to be. The assessment can be made more rigorous by adding aversive consequences (e.g., electric shock) of the same lever pressing, and seeing how much self-inflicted punishment the animal will endure while continuing to work for drug injection.

An indirect method consists of administering a test drug to an animal in one distinctive environment and a placebo in another, and then seeing which environment the animal prefers when given a free choice. Preference for the environment previously paired with the drug implies that the drug is reinforcing; preference for the other environment, that the drug is aversive. The direct and indirect methods do not give identical results. In general, however, any drug that is self-administered by experimental animals is also self-administered by humans, but humans will take some drugs that animals will not.

Types of reinforcement. The discrepancy noted above is probably due to the fact that self-administration studies often explore only primary positive reinforcement resulting from a direct pharmacologic effect on a central neuronal circuit that is presumed to mediate reinforcement. In humans, however, other forms of reinforcement probably play a major role. For example, secondary positive reinforcement can occur when drug taking is linked to another reinforcer such as peer group approval for using the drug. Negative reinforcement occurs when the drug taking results in alleviation of an unpleasant or aversive state such as anxiety, frustration, pain, or depression. These types of reinforcement are more difficult to study in experimental animals, but it is possible to do so.

Stimulus control. One further factor is required to convert strongly reinforced drug taking into addiction. Drug taking does not occur in a vacuum, but in a social, physical, and psychological context. If the same context is present often enough, cues derived from it become conditionally linked to the drug taking, which thus becomes a conditioned response to those cues. In operant conditioning terms, the cues set the occasion for the drug taking response. In classical conditioning terms, the cues evoke the response, which is therefore no longer a "voluntary" action. The subject is then considered to be addicted.

Substitution of dependence. A useful, rapid technique for screening new drugs that may carry dependence liability is to test their ability to substitute for another drug on which the subject (human or experimental animal) is already dependent. If an animal trained to self-administer cocaine, for example, continues lever pressing when a different drug replaces cocaine, the new drug is presumably reinforcing and potentially addictive. If a subject with physical dependence on morphine fails to show a withdrawal reaction when a new drug replaces the morphine, the new drug is taken to have morphine-like actions and risks. However, physical dependence can be produced by passive exposure (i.e., without self-administration), and substitution of physical dependence is therefore a less valid measure of addictive properties.

Significance of tolerance and physical dependence

Chronic ingestion of a drug, with sufficient frequency and amount, gives rise to adaptive changes in neuronal biochemistry and physiology which, augmented by behavioral factors including learning and classical conditioning, offset the action of the drug and thus give rise to tolerance. When the drug is withdrawn, these changes are maladaptive, giving rise to the withdrawal reaction. The underlying neuronal changes vary by drug and include altered release of various neurotransmitters, changes in binding properties of drug receptors, increases in activity of adenylate cyclase, (Na^+ + K^+)-adenosine triphosphatase and other membrane-bound enzymes, and changes in lipid composition of plasma membranes. It is not yet known which of these are actually mechanisms of tolerance and which are simply manifestations of it.

The importance to addiction is that tolerance and physical dependence can increase the preponderance of reinforcing over aversive effects of the drug. Tolerance to the aversive effects of drugs appears to be much greater than to the reinforcing or discriminative effects. Physical dependence can give rise to a new source of negative reinforcement as users learn to treat their withdrawal symptoms by taking more drugs. Thus, tolerance and physical dependence can increase the strength of the addiction.

Drugs and mechanisms of reinforcement

Drugs can be ranked in a hierarchy of intrinsic (primary) reinforcing efficacy. In experimental animals, cocaine and amphetamine-like drugs appear to be strongest, followed by heroin and other potent μ-agonist opioids. Anxiolytics, rapidly acting barbiturates, and other sedatives are moderately effective, but ethanol is only weakly and inconsistently so. Cannabis and hallucinogens are aversive rather than reinforcing in animal models; their use by humans implies an important role of secondary or negative reinforcement.

The route of administration is very important. Operant conditioning requires a close temporal relation between the drug-taking behavior and the onset of reinforcing effects. Therefore intravenous injection, resulting in almost immediate onset, has the highest addictive risk; inhalation or smoking is almost as rapid and effective; oral ingestion of the same drug, with a long absorption delay, is the least effective.

Various hypotheses concerning the neural substrates of reinforcement and addiction have been proposed. The strongest evidence at present implicates dopaminergic or noradrenergic fibers in the ventral tegmentum, and a possible modulatory role of endogenous opioid peptides, but the evidence is too incomplete to permit a clear picture. It was recently reported (Bozarth and Wise, 1984) that the reinforcing action of morphine is exerted in the ventral tegmentum whereas physical dependence results from action in the periventricular gray substance.

Individual predisposition

Innumerable attempts to define an "addiction-prone personality" have been unsuccessful. However, there is good evidence for a genetic predisposition to alcoholism in the sons of alcoholic fathers. It is not known whether the inherited trait is a greater drug sensitivity of the primary reinforcement mechanism, or a feature such as an affective disorder that increases the likelihood of secondary or negative reinforcement by alcohol. There is as yet no strong evidence concerning similar genetic factors in addiction to other drugs. Nonhereditary psychological or experimental factors probably also modify the magnitude of reinforcing effects of a drug in different individuals.

Socioeconomic and cultural factors

Epidemiological studies indicate that average values for daily consumption of alcohol by individuals within a population are distributed along a unimodal curve that shows no recognizable discontinuity between alcoholics and social drinkers. Preliminary data suggest similar distributions for consumption of other potentially addictive drugs, including illicit ones. Changes in mean per capita consumption by the whole population, in response to economic, legal, and other factors, are accompanied by corresponding shifts in the entire distribution-of-consumption curve. In one experiment, alcoholics and nonalcoholics showed the same proportional change in alcohol intake in response to price change, even though their absolute intake levels were very different. Such effects are also superimposed on cultural factors which give rise to differences in per capita intake in different societies. Thus, it appears that socioeconomic and cultural factors may affect the probability of trying a drug and experiencing its reinforcing effects, but they do not explain why one individual becomes addicted while another does not.

Further reading

Bozarth MA, Wise RA (1984): Anatomically distinct opiate receptor fields mediate reward and physical dependence. *Science* 224:516–517

Fishman J, ed. (1978): *The Bases of Addiction*. Berlin: Dahlem Konferenzen

Moore MH, Gerstein DR, eds. (1981): *Alcohol and Public Policy: Beyond the Shadow of Prohibition*. Washington DC: National Academy Press

Smith JE, Lane JD, eds. (1983): *The Neurobiology of Opiate Reward Processes*. Amsterdam: Elsevier

Woolverton WL, Schuster CR (1983): Behavioral and pharmacological aspects of opioid dependence: mixed agonist-antagonists. *Pharmacol Rev* 35:33–52

Aging of the Brain

Arnold B. Scheibel

The processes of maturation and aging of the brain are becoming increasingly relevant and active research areas as the average longevity of the population increases. In 1900, about 4% of the population of the United States exceeded the age of 65. In 1980, the figure approximated 12%, and it is predicted that by the year 2000, more than 15% of the population (at least 35 million people) will be 65 years of age or older. Although remarkably little is known about the aging process, it is increasingly clear that senility and aging are not synonymous. One of the major thrusts of recent neural research has been to separate the phenomenon of normal, vigorous aging from a broad range of disease patterns that alter the structure and function of the brain and the cognitive and psychosocial behavior of the individual.

Possible causal factors

Experience indicates that all living things have finite lives, whether they be the ephemeral summer of the butterfly, a thousand days for the rat, the reported hundred-year-long lives of large tortoises, or the multi-thousand-year histories of giant redwoods and bristle cone pines. The concept of genetic programming of the aging process seemed to receive support from Hayflick's studies of the late 1950s and early 1960s in which young, actively dividing connective tissue cells raised in vitro appeared limited to a certain maximum number of mitotic divisions (50), after which degenerative changes invariably set in. In retrospect, these classic experiments now seem less convincing due to the methodological errors involved, and, although the concept of genetic control remains intuitively appealing, the case is considered far from proven. A presently favored alternative theory operating either independently, or more likely in tandem with the proposed genetic constraint, envisages the possibility of progressive damage to the cell from external or internal factors, resulting in a cumulative pattern of dysfunctions. For example, toxic factors within the environment (i.e., chemical carcinogens and background radiation of terrestrial or galactic origin) may slowly affect the cell DNA content and protein-synthesizing machinery, leading to errors in synthesis with resultant progressive changes in cell structure and function. This summation of many small developing errors in the synthetic and enzymatic machinery of the cell is conceived as mounting to a point beyond which conditions for cell life become impossible (error-catastrophe theory). Obviously, this putative process might complement and potentiate a set of existing genetic instructions also directed toward the cell's eventual demise.

However, normal oxidative metabolic activity of the cell may itself result in cumulative damage. Of particular interest is the development of a family of free radicals of oxygen (e.g., singlet oxygen, hydrogen peroxide, superoxide, and hydroxyl radicals) that, through cross-linkage or cleavage reactions, may permanently alter the structure of the cell. These ideas are still in an early stage of development and exemplify the questions that surround the processes of aging in general, and of the nervous system in particular.

Structural changes

A number of changes, both gross and microscopic, have been reported in the brains of the aged, although variation is the rule rather than the exception. The brain itself is often somewhat reduced in size and weight, especially beyond the eighth decade of life. Ten to fifteen percent decrements are frequently quoted, although interindividual differences are large. Some aged brains show mild to moderate gyral atrophy and sulcal widening, but this is not the rule. The surrounding meninges are frequently more opaque and milky in appearance than those of the young brain, and may be adherent to underlying cortex.

Isolated deposits of calcium are found in and around the Pacchionian granulations near the vertex of the hemisphere. The ventricular system may show alterations in silhouette, and computed axial tomography (CT) and nuclear magnetic resonance (NMR) scans may show enlarged ventricular shadows and evidence of cortical atrophy. Such phenomena are noted with increasing regularity in patients over age 80.

Microscopically, a group of well-known stigmata are seen in routine Nissl or reduced silver-stained preparations. Their incidence varies rather widely among individuals, although there is a general tendency toward increased frequency with age. Neuronal cell bodies gradually accumulate masses of refractile granules with high lipid content, known as lipofuscins. The significance of these so-called aging pigments is not clear, and they have been found as early as the 10th year of life in cells of the inferior olive. The present consensus is that they represent the remains of lysosomal and mitrochondrial membranes that have accumulated, due perhaps to gradual failure of mechanisms for turnover and reutilization. As such, they do not necessarily constitute a direct threat to the neuron except to the extent that they preempt increasing amounts of cytoplasmic space previously used for synthesis of glycoproteins, lipoproteins, neurotransmitters, etc. One interesting and opposing view maintains that the appreciable content of myoglobin and respiratory enzymes allows lipofuscin granules to serve a positive role in providing energy to neurons under conditions of low oxygen tension. Coarser vacuolization of the cytoplasm, usually concentrated in the area of development of the apical dendrite shaft in cortical pyramidal cells, is known as granulovacuolar degeneration of Simchowitz and is of equally unknown etiology.

The neurofibrillary tangle is a structural alteration of neuronal cytoplasm that may involve soma, dendrites, and axon. Initially recognized by Alzheimer and Simchowitz as one of two defining microcriteria of individuals dying with dementia, it is also found in very small numbers in the normal aged. Electron microscope study reveals that the tangle consists of paired helical filaments about 22 nm in width which braid loosely around each other with a characteristic series of partial twists, each 80 nm from the next. Of uncertain origin, these paired filaments do not seem directly derivative of the neurofilaments and microtubules they replace. However, there is some evidence for cross-reaction between antibodies to tubulin (substrate for microtubules) and neurofibrillary tangles, but not to neurofilaments. The import of the neurofibrillary tangles to the neuron remains uncertain, but it has been suggested that their development impedes intracellular axonal and dendritic transport.

Accompanying these characteristically intracellular alterations is the development of small foci of destruction in the surrounding neuropil, the senile plaque of Alzheimer. These are classically described as containing central cores of Congo Red–positive, amyloid-like material, surrounded by radial aurae of degenerating dendritic and axonal tissue. They constitute the second component of the two major histopathological criteria of dementia and, like neurofibrillary tangles, are present in more restricted number in the brains of most aged individuals. The density of senile plaques has been said to correlate positively with the degree of cognitive impairment in dementia, although these data have recently been subject to question.

Intensive biochemical and immunocytological analyses of these structures are presently under way. Now known to consist of an amyloid type-B core, rich in immunoglobulin G (IgG), it has been suggested that this primarily represents an antigen-antibody complex, perhaps resulting from globulin leakage through a failing blood brain barrier with subsequent attack upon neurons, erroneously identified as foreign antigens. Indeed the high incidence of perivascular locations for plaques and the report from several laboratories of high titers of anti-brain antibodies in old mice support this interpretation. On the other hand, data have also been presented identifying these areas as degenerating cholinergic presynaptic terminals. This interpretation has attracted a good deal of recent interest because of the developing age- and dementia-related deficit in cholinergic neurons in the basal forebrain (especially the nucleus basalis of Meynert). However, the size and structure of senile plaques do not particularly support this idea.

All the histopathological changes so far described appear most intensively in the limbic system, especially the entorhinal cortex and hippocampus. They are also found widely throughout the rest of the central nervous system, however, including cerebral neocortex, diencephalon, brain stem, and spinal cord. Since interindividual variation is the rule in the aged brain, these descriptions represent a distillation and summary rather than the expected picture of the brain of each aged individual.

The problem of neuronal loss has been a hotly debated one and is still not settled. The few quantitative studies dating from the late 1950s and 1960s suggested neuron loss of up to 30% in the normal aging cerebral neocortex. More detailed recent investigations based on larger numbers of cases indicate that these levels of neuronal loss, in normal aging at least, may have been excessive, based as they were on the study of a few areas in a limited number of brains. More broadly sampled data suggest a pattern of modest loss of neurons, primarily large cortical cells, with compensatory dendritic growth in adjacent neurons. The cholinergic cell masses of the basal forebrain and the noradrenergic cells of the locus ceruleus undergo undoubted change during the aging process, but even here, declining function may be more an expression of waning metabolic vigor and decreased axodendritic dimensions than of massive neuronal loss. This is in sharp contrast to dementing syndromes such as Alzheimer's disease where more than half of the complement of locus ceruleus cells may disappear.

Carefully controlled studies indicate that in the rat the only significant cerebral cell losses occur during the first 100 days of life in what might be considered a period of adaptation and fine tuning to the environment. After this, there is no further discernible neuronal loss, even at 900 days of age when the rats are beginning to die of natural causes.

Although the weight of evidence from both animal and human studies now increasingly points to negligible neuronal loss during the normal, uncomplicated aging process, a large number of cells appear to undergo changes in the dendritic (and axonal) extensions of the soma. Many neurons show progressive restriction and atrophy of their more peripheral dendrite branches and, especially in cortical pyramidal cells, among the basilar shafts. Accompanying this is irregular loss of dendrite spines and frequent beaded swellings along the remaining dendritic branches. These changes can be related in general terms to progressive loss of protein-synthesizing capabilities due, perhaps in part, to increasing incursions upon cytoplasmic space by lipofuscin deposits and neurofibrillary tangles. However, it has also been shown that the potential for neuronal growth is not lost during aging. Accompanying the progressive destruction of some dendritic systems, it appears that other neurons grow further dendritic extensions, thereby increasing their available synaptic areas (Fig. 1). The concept of two types of neuronal response to aging, one involving dendritic retraction and one reactive dendritic expansion, brings with it a number of exciting implications for providing more effective and fulfilling lives for the elderly.

During the first half of this century, brain aging and dementia

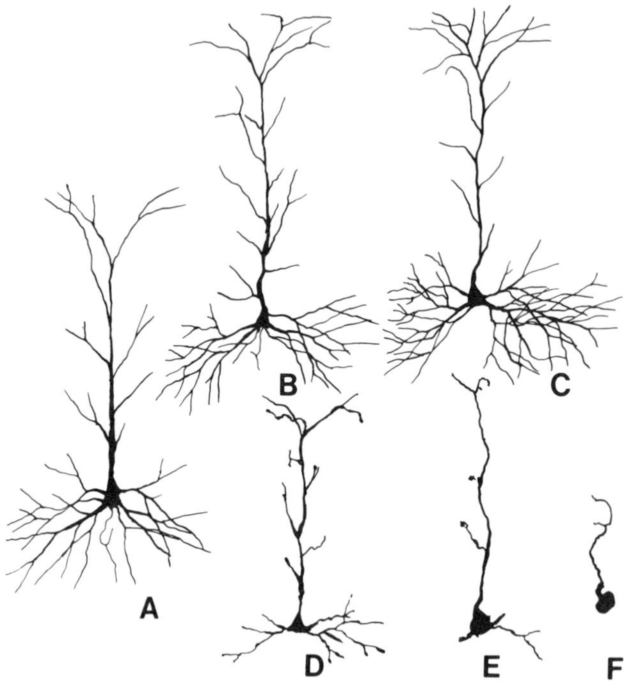

Figure 1. Two possible patterns of age-related alterations in cortical pyramidal cells. The normal mature neuron (A) may show regressive dendritic changes characterized by loss of basilar dendritic branches and eventual loss of the entire dendritic tree (D, E, F). Other neurons (B, C) may show progressive increase in dendritic branching. Drawing based on Golgi impregnations.

were usually associated with visible changes in the major arteries of the central nervous system. Cerebral arteriosclerosis was considered a necessary accompaniment and, in fact, virtually synonymous with aging and senility. Careful studies during the 1950s showed the inaccuracy of such concepts, and today, large-vessel disease is seldom considered to contribute significantly to the picture of general brain aging. Recent investigations with positron emission tomography (PET) scanning methods and xenon clearance techniques have again called attention to the adequacy of blood flow and oxygen and glucose metabolism in the aging brain. The scanning electron microscope and immunohistochemical studies have, coincidentally, focused interest on the microcirculation (capillary bed) of the brain and on the plexus of neural fibers that normally innervate their walls. Subtle changes in this delicate but all-pervasive system may prove to have significant impact on the maintained vigor or decline of brain structure and function.

Biochemical and metabolic changes

It is becoming clear that the aging process entails a broad range of biochemical and metabolic changes. Among these, alterations in neurotransmitter content and activity figure prominently. As many as 90% of neurons are cholinergic, synthesizing, transporting, and releasing, or else being synaptically dependent upon, acetylcholine. Cholinergic systems are highly energy dependent, and age-related decreases in several important glycolytic enzymes have been reported. A link might thus conceivably be postulated between altered glycolytic energy mechanisms and cognitive function.

The most important concentration of acetylcholine-rich neurons is found in the basal forebrain area, in and around the nucleus basalis of Meynert and ventral pallidum. Most studies indicate mild to moderate age-related cell loss in these areas. Associated with (although not entirely dependent upon) this loss are decreases in acetylcholine content in various portions of the brain, lower levels of the synthesizing enzyme, choline acetyltransferase, and decreased numbers of acetylcholine receptor binding sites. The relevance of these alterations to cognitive changes in the elderly, such as impairment of recent memory function, are not entirely clear. Of uncertain import, also, is the apparent loss of fluidity of the cytoplasmic membrane of the cell as the lipoprotein structure changes, a possible result of progressive diminution of choline content. Age-related increases in membrane microviscosity may bring with them a significant train of sequelae including, initially, a higher capacity for receptor binding followed by enhanced rates of receptor loss with eventual overall reduction in receptor binding affinity. Mechanisms responsible for these changes remain uncertain, but it has been postulated that increased microviscosity of the cell membrane leads first to increased exposure of those receptors out of the plane of the membrane, after which the uncovered receptors are progressively sloughed off into the surrounding medium. This suggests that membrane-bound proteins such as receptors are maintained in dynamic equilibrium by the extent of fluidity of the surrounding membrane lipids. Attempts to forestall or modify age-related losses in membrane fluidity through administration of active lipid fractions are presently under investigation.

Catecholaminergic systems also show moderate to marked alteration in the aged brain. Levels of both dopamine and norepiniphrine synthesis decrease, as do the numbers of adrenergic and dopaminergic receptors. Marked attenuation of norepinephrine synthesis in the hypothalamus may, in turn, be responsible, at least in part, for decrease in the synthesis of certain hypothalamic hormones.

Significant cell loss has been reported in the locus ceruleus, the single most important source of brain parenchymal norepinephrine. The substantia nigra, a major source of dopamine, also undergoes histological alterations that include microvascular changes and neuronal loss. The serotonin-rich systems of the raphe seem, on the other hand, to be somewhat less vulnerable. Resultant maintenance of minimally disturbed titers of serotonin in a setting of falling catecholaminergic levels may conceivably be related to a high incidence of disturbed sleep patterns and depression in the aged.

A broad range of age-related neuroendocrine changes are known to exist. Although such alterations are documented by innumerable anecdotal observations, regional study of the mechanisms involved is in its early phases. For instance, the importance of the ovary as a pacing factor in the aging female rat is receiving documentation in many laboratories. At an equally obvious level, the high likelihood of occurrence of benign prostatic hypertrophy and prostatic cancer in the aging male is being related in meaningful fashion to age-related alterations in testosterone metabolism that may lead to unbalanced growth stimulation. In a more global sense, the entire aging process can apparently be slowed by early restriction of nutritional input, which seems to delay pubescence and maturation. Finally, increasing imbalance among the various neurotransmitters, and in particular among those that are aminergic, appears to exert progressive impact upon the hypothalamus, pituitary, and pineal gland, leading to cascades of dysfunction that manifest themselves at every level of organization.

Although the picture presented appears to emphasize regressive elements, a much more positive picture of the normal aging brain can now be painted. Accumulated experience can more than make up for age-related loss in the speed of cerebral processing or of memory recall. Experimental evidence conclusively demonstrates the continued plasticity of even the very old brain. Continual environmental enrichment and challenge appear to enhance cognitive power in the aged. Under conditions of optimally maintained health, increasing numbers of individuals in their 70s and 80s and beyond continue to hold responsible positions, or otherwise distinguish themselves in the arts and in certain of the sciences. In every case, the touchstone appears to be activity, involvement, and purpose!

Further reading

Finch C, Hayflick L, eds (1977): *Handbook of the Biology of Aging*. New York: Van Nostrand-Reinhold

Stein DG, ed (1980): *The Psychobiology of Aging: Problems and Perspectives*. New York: Elsevier, North Holland

Diamond MC, Connor JR (1981): A search for the potential of the aging cortex. In: *Brain Neurotransmitters and Receptors in Aging and Age Related Disorders*, Enna SJ, Samorajski T, Beer B, eds *Aging*, 17:43–58. New York: Raven Press

Alcoholism

Donald W. Goodwin

Alcoholism refers to excessive consumption of alcoholic beverages resulting in persistent social, psychological, and medical problems. Prevalence rates differ from country to country, and the true prevalence is not known. In the United States and northern European countries, it is estimated that 3–5% of men and 0.1–1% of women can be described as alcoholic at some time in their lives. "Alcohol dependence" is synonymous with alcoholism. "Problem drinking" and "alcohol abuse" are other terms which overlap with alcoholism, although they usually include larger numbers of individuals whose problems from alcohol are less severe and persistent.

Clinical picture

Alcoholism is a behavioral disorder. The specific behavior that causes problems is the consumption of large quantities of alcohol on repeated occasions. The motivation underlying this behavior often is obscure. When asked why they drink excessively, alcoholics occasionally attribute their drinking to a particular mood, such as depression or anxiety, or to situational problems. They sometimes describe an overpowering need to drink, variously described as a craving or compulsion. Just as often, however, alcoholics are unable to give a plausible explanation for their excessive drinking.

Like other drug dependencies, alcoholism is accompanied by a preoccupation with obtaining the drug in quantities sufficient to produce intoxification over long periods. Early in the course of alcoholism the patient may deny this preoccupation. As part of this denial or rationalization, alcoholics tend to spend their time with other heavy drinkers. As problems from drinking become more serious, alcoholics may drink alone, sneak drinks, hide the bottles, and take other measures to conceal the seriousness of the condition. This is accompanied by feelings of remorse which in turn may produce more drinking. Remorse may be particularly intense in the morning, when the alcoholic has not had a drink for a number of hours. This may provoke morning drinking.

Prolonged drinking, even if initially guilt and anxiety relieving, also produces anxiety and depression. The full range of symptoms associated with depression and anxiety neurosis—insomnia, low mood, irritability, and anxiety attacks—occurs. Alcohol temporarily relieves these symptoms, resulting in a vicious cycle of drinking-depression-drinking which ultimately may result in an alcohol withdrawal syndrome. Often the patient tries to stop drinking and may succeed for several days or weeks, only to relapse.

Repeated relapses lead to feelings of despair and hopelessness. By the time the patient consults a physician, he has often hit "bottom." His problems have become so numerous that he feels nothing can be done about them. He now acknowledges his alcoholism, but feels powerless to change.

Blackouts, amnesia for events that occur while drinking, are particularly distressful, since these drinkers may fear they have unknowingly harmed someone or behaved imprudently while intoxicated. The amnesia is anterograde: remote and immediate memory are intact, but there is a specific short-term memory deficit, a situation that permits complicated acts which may appear normal to a casual observer.

In addition to psychological symptoms, alcoholics commonly have social and medical problems from drinking. Alcoholics have a high rate of marital separation and divorce. They often have work troubles, including frequent absenteeism and job loss. They also have a high frequency of accidents—in the home, on the job, and in cars. Of the more than 30,000 highway fatalities each year in the United States, about half involve a driver who has been drinking, usually an alcoholic. Nearly half of convicted felons are alcoholic and about half of police activities in large cities are associated with alcohol-related offenses.

Medical problems fall into three categories: (1) acute effects of heavy drinking; (2) chronic effects; and (3) withdrawal effects. Rapid consumption of large amounts of alcohol can cause death by depressing the respiratory center in the medulla. Acute hemorrhagic pancreatitis occasionally occurs from a single heavy drinking episode. Nearly every organ system can be affected. Gastritis and diarrhea are common. The most serious effect of alcohol on the gastrointestinal tract is liver damage. A single large dose of alcohol reversibly increases the fat content of the liver. The connection between this and cirrhosis, however, is unclear. Most patients with portal (Laennec's) cirrhosis, in Western countries, are excessive drinkers, but less than 10% of alcoholics develop cirrhosis.

Alcohol damages the nervous system by causing vitamin deficiencies. Whether alcohol itself is a neurotoxin is less certain. Peripheral neuropathy, the most common neurological complication, apparently results from multiple vitamin B deficiencies and usually is reversible with adequate nutrition. Retrobulbar neuropathy may lead to amblyopia (sometimes called tobacco-alcohol amblyopia), also usually reversible with vitamin therapy.

Other neurological complications include anterior lobe cerebellar degenerative disease and the Wernicke-Korsakoff syndrome. The latter results from thiamine deficiency. The acute Wernicke stage consists of ocular disturbances (nystagmus of sixth nerve palsy), ataxia, and confusion. It usually clears in a few days but may progress to a chronic brain syndrome (Korsakoff psychosis). Short-term memory loss (anterograde amnesia) is the most characteristic feature of Korsakoff's psychosis. The Wernicke-Korsakoff syndrome is associated with necrotic lesions of the mammillary bodies, thalamus, and other brain stem areas. Thiamine corrects early Wernicke signs rapidly and may prevent development of an irreversible Korsakoff dementia.

Other medical complications of alcoholism include cardiomyopathy, thrombocytopenia, anemia, and myopathy.

The term alcohol withdrawal syndrome is preferable to delirium tremens (DTs). The latter refers to a specific manifestation of the syndrome. The most common withdrawal symptom is tremulousness, which usually occurs a few hours after cessation of drinking and may begin while the person is still drinking ("relative abstinence"). Transitory hallucinations also may occur. If so, they usually begin 12 to 24 hours after drinking stops. Grand mal convulsions ("rum fits") occur occasionally, sometimes as long as two or three days after drinking stops. As a rule, alcoholics experiencing convulsions do not have epilepsy; they have normal electroencephalograms when not drinking and experience convulsions only during withdrawal. Delirium tremens is infrequent, usually occurring in alcoholics who are medically ill. A severe memory disturbance is required for the diagnosis. Delirium tremens usually begins two or three days after drinking stops and subsides within one to five days, or when the patient recovers from the medical illness.

Suicide is an important complication of alcoholism. About one-quarter of suicides are alcoholic, mainly men over age 35.

Natural history

The natural history of alcoholism is somewhat different in men and women. In men the onset usually is in the late teens or twenties. The course is insidious. The first hospitalization usually occurs in the late thirties or forties. Symptoms of alcoholism in men rarely occur for the first time after age 45. Alcoholism has a higher spontaneous remission rate in men than is often recognized. The incidence of first admissions to psychiatric hospitals for alcoholism drops markedly in the sixth and seventh decades of life. Although the mortality rate among alcoholics is two to three times that of nonalcoholics, this is probably insufficient to account for the apparent decrease in problem drinking in middle and late middle life. The course of the disorder is more variable in women. The illness often begins later in life, and spontaneous remission appears to be less frequent.

Treatment for alcoholism is generally considered to be unsatisfactory. Approaches include psychotherapy (individual and group), behavioral therapy, aversive therapy (using emetics), and pharmacotherapy (antidepressant drugs, anxiolytic drugs, lithium, and disulfiram). There is little or no evidence that any of these treatments alters the natural history of the illness, that is, produces improvement rates superior to those which occur without treatment. The value of treatment, however, remains an open question. There have been very few controlled studies with patients randomly assigned to treatment versus no-treatment groups. Some studies of disulfiram and lithium indicate modest effectiveness; others do not. (Disulfiram deters drinking by causing deleterious effects combined with alcohol; lithium is a treatment for mania.) There is no evidence that prolonged hospitalization produces a better outcome than short-term outpatient treatment.

Etiology

About 70% of adults in most Western countries drink alcohol. Of these, about one in ten are addicted, i.e., drink large amounts on repeated occasions with adverse consequences. It is not known why a small minority of drinkers become alcoholic while most are relatively unharmed. The development of alcoholism, however, is influenced by three factors which may have causal importance: (1) genetic factors; (2) ethnicity; and (3) childhood behavior problems.

Alcoholism runs in families, and does so even when the children of alcoholics are separated from their parents and raised by nonalcoholic adoptive parents. Twin studies further suggest genetic factors. Monozygotic twins are more often concordant for alcoholism than dizygotic twins. A family history of alcoholism constitutes the strongest risk factor for the development of alcoholism. About 20% of sons and 5% of daughters become alcoholic compared to rates of alcoholism in the general population of 5% in men and 1% in women. Many alcoholics do not have alcoholism in the family, and there has been a recent trend toward separating alcoholics into familial and nonfamilial types. The familial type apparently has a younger age of onset and a more severe course. Other studies compare children of alcoholics with children of nonalcoholics. Members of the former group more often have brain abnormalities and a history of childhood hyperactivity.

Alcoholism is unevenly distributed among national and ethnic groups. Ireland, the Soviet Union, and the Baltic nations have a high rate, and countries along the Mediterranean Basin a low rate, except for France (which has the world's highest cirrhosis rate). There may be a correlation between parental attitudes toward drinking and subsequent alcoholism. Ethnic groups which condone drinking in moderation but condemn drunkenness appear to have a low rate of alcoholism (e.g., Jews). Childhood behavior problems weakly predict future alcoholism.

No personality type is associated with alcoholism, except the antisocial personality. The latter group has a high rate of alcoholism. Prospective studies have been of little hope in predicting who will become alcoholic. Prealcoholic teenagers, if anything, appear more stable and self-confident than their peers. An unstable family life apparently does not increase the risk of alcoholism unless the instability is produced by an alcoholic parent, in which case the increased risk of alcoholism in the children may be a function of genetic factors.

Millions of individuals are "protected" from becoming alcoholic because of a profound intolerance for alcohol. Orientals are particularly susceptible to adverse physiological effects from small amounts of alcohol, which may explain the low rate of alcoholism in the Orient. More Caucasian women than men are physiologically intolerant of alcohol. The Orientals' intolerance appears related to atypical forms of alcohol dehydrogenase and aldehyde dehydrogenase which favor an accumulation of acetaldehyde, producing effects similar to those occurring in alcohol-disulfiram reactions.

Neurochemical theories

Two lines of research have emerged which eventually may help explain tolerance, physical dependence, and alcohol-induced euphoria. One involves biological membranes. Chronic consumption of alcohol induces an adaptation of membrane phospholipid composition, causing increased membrane rigidity (decreased fluidity). The increased rigidity impairs normal membrane function and may contribute to tolerance. In the restored presence of moderate concentrations of alcohol the membrane becomes sufficiently fluid to resemble normal membranes, and this may partly explain physical dependence. The evidence for these changes, however, is inconsistent and further research is needed before the link between the physical and functional effects of ethanol can be understood.

Another promising line of research began with the observation in 1970 that biogenic amines and their aldehyde metabolites form condensation products with acetaldehyde, the first metabolite of alcohol. These products structurally resemble opiate alkaloids, leading to speculation that addiction to alcohol and opiates might have a final common biochemical pathway. The condensation products are formed in minute quantities but bind to certain opiate receptors and otherwise behave in some ways like opiates. Naloxone, an opiate antagonist, partially blocks some effects of alcohol (e.g., incoordination) and has been reported anecdotally to reverse alcohol-induced coma. Also, injection of the products generically called tetrahydroisoquinolines (or TIQs) into rat and monkey brains has resulted in excessive consumption of alcohol by these animals, which persisted long after the substances had been injected. Finally, TIQs have been found in cerebral spinal fluid (CSF) of alcoholics and nonalcoholics after drinking alcohol, with higher quantities in the cerebrospinal fluid of alcoholics.

TIQ research also suffers from many inconsistencies but has stimulated research directed at finding a neurochemical basis for variations in alcohol ingestion by animals and humans.

Further reading

Vaillant G (1983): *The Natural History of Alcoholism.* Cambridge: Harvard University Press

Goodwin D (1981): *Alcoholism: The Facts.* Oxford: Oxford University Press

Pattison EM, Kaufman E, eds (1982): *Encyclopedic Handbook of Alcoholism.* New York: Gardner Press

Alzheimer's Disease

Joseph T. Coyle

Alzheimer's disease (AD) and senile dementia of the Alzheimer's type, a distinction based primarily on age of onset, appear to be a similar, if not the same, disease process. It is now apparent that AD accounts for 40% to 60% of all cases of dementia with onset in adulthood and is two to threefold more frequent than multi-infarct dementia. It is estimated that 3–5% of individuals over the age of 65 suffer from AD, and current costs for nursing home care in the United States for patients afflicted with AD exceed $10 billion per year.

Clinical manifestations

The most frequent initial symptom of AD is deterioration in recent memory with relative preservation of remote memory. However, with progression, a global deterioration in memory occurs, resulting in loss of recognition of even family members. Next to the memory deficits, aphasias (impairments in verbal expression) are a frequent symptom in the early stages of AD; this progresses to difficulties in reading and in written language, and as the dementia progresses, neologisms, nonsense words, and echolalia appear. Apraxia, the inability to perform complex routine motor acts, also occurs, resulting in confusion on how to use familiar objects such as dinnerware, tools, and writing instruments. Deterioration in this realm leaves the individual incapable of performing the most basic activities of daily living such as dressing and feeding. Notably, impairments in the level of consciousness do not occur until the end stages, and the disorder is not associated with focal, motor, or sensory neurological symptoms such as paralysis or anesthesia, although primitive reflexes may emerge. Ultimately, the AD patient, from 5 to 15 years after the onset of symptoms, is left bedridden and mute, usually dying from intervening infections.

Neurohistopathology

Unequivocal diagnosis of AD requires the histopathologic demonstration of the characteristic stigmata of the disorder in the brain. These alterations include neuritic plaques, neurofibrillary tangles, and granulovacuolar degeneration. Neuritic plaques are composed of extracellular deposit in the neuropil of amyloid, which consists of beta pleated sheets of protein that exhibit birefringence under polarized illumination. The source of amyloid protein is unknown. Surrounding the amyloid core, especially in "immature plaques," are dystrophic neurites, visible with silver stains, that often have clublike terminations resembling growth cones. Neuritic plaques, which also occur at a density one magnitude lower in the normal human aged brain, are found predominantly in the cortex, hippocampus, and limbic regions in AD. The density of neuritic plaques in cerebral cortex correlates significantly with the degree of cognitive impairments seen in patients before death. Neurofibrillary tangles represent the accumulation of highly cross-linked protein fibrils within neuronal cell bodies. The neurofibrillary tangles are remarkably resistant to conditions that readily solubolize most proteins. Whether neurofibrillary tangles represent proteins unique to AD remains uncertain, although recent studies have revealed immunocross-reactivity between a component of the neurofibrillary tangles and neurofibrils. Neurofibrillary tangles are concentrated in cortical, hippocampal, and limbic neurons but are also found in neuronal perikarya within the reticular core. Granulovacuolar degeneration represents the accumulation of silver-staining vacuoles that distend neuronal cell bodies, especially the hippocampal pyramidal cells.

It is unclear whether neuronal cell loss occurs in the cerebral cortex in AD and, if so, to what extent. In part, this uncertainty derives from the fact that normal *aging* is associated with cortical atrophy and neuronal cell loss. Recent studies using cell counting techniques have not revealed striking reductions in cell number in cortex when AD patients are compared to age-matched controls; however, in some cortical regions, significant reductions in the number of the largest cells, presumably neurons, have been noted. These studies, which are based on Nissl stain techniques, cannot address the structural integrity of neurons, and limited results from Golgi analysis suggest a marked disruption in neuronal cell structure with a high incidence of irregular, misshapen neurons with a ballooned appearance in AD.

Synaptic neurochemical pathology

A consistent finding in postmortem analysis of brains from individuals who have died with pathologically confirmed AD is a striking reduction in presynaptic markers for cholinergic neurons, including choline acetyltransferase and acetylcholinesterase, in the cerebral cortex, hippocampal formation, and the limbic system. In analysis of biopsies of cortex in AD, the high-affinity uptake process for choline and the ability to synthesize acetylcholine has been found to be similarly reduced. The predominant source of cholinergic innervation to these structures is derived from neuronal perikarya located in the basal forebrain magnocellular complex, including the nucleus basalis of Meynert, the diagonal band of Broca, and the medial septal nucleus. Histopathologic studies have generally indicated that there is significant loss of these cholinergic perikarya or that they are shrunken in appearance in AD. The critical role of cholinergic forebrain projections in the pathophysiology of AD has been further supported by the findings that the severity of reduction of cholinergic markers correlates significantly with the degree of cognitive impairments in patients before death and that the dystrophic neurites in the neuritic plaques, an essential pathologic feature of AD, stain for acetylcholinesterase and choline acetyltransferase, two markers for cortical cholinergic terminals. However, the severity of cortical and hippocampal cholinergic deficits does not appear to correspond closely with the cell loss in the basal forebrain cholinergic nuclei. Thus, there is reason to believe that the proximate cause of the cortical cholinergic deficits is not due to an anterograde degeneration but rather reflects a retrograde degeneration of the cholinergic projections, with perikaryal destruction as a terminal event.

The forebrain cholinergic projections are not the only components of the reticular core that are affected in AD. Variable reductions in presynaptic markers for the noradrenergic neurons have also been noted, and histological studies of the nucleus locus ceruleus indicate that major neuronal loss occurs primarily in patients with earlier age of onset of AD. Similarly, variable reductions in the presynaptic markers for serotonergic neurons in cortex and hippocampus have also been observed in AD. Two neuropeptides—substance P and somatostatin—are reduced in concentration in the cortex and hippocampus in AD. Whether these deficits reflect a degeneration of neurons containing these neuropeptides or an alteration in their synthesis or degradation is unknown.

Markers for a variety of neurons located within the cerebral cortex and hippocampus have also been assessed. Notably, muscarinic acetylcholine receptors, which are presumably concentrated on the cortical neurons receiving cholinergic input, are not reduced in AD. Furthermore, presynaptic markers for GABAergic (gamma-aminobutyric-acidergic) neurons, which are known to be local circuit neurons within the cortex and hippocampus, are unaffected or mildly reduced. In addition, the concentrations of vasoactive intestinal peptide, cholecystokinin, and enkaphalin, neuropeptides localized in neurons intrinsic to the cerebral cortex, also remain in the normal range in AD. Thus, the synaptic neurochemical studies point to rather selective alterations in the integrity of components of the reticular core projections to the telencephalon, especially the cholinergic neurons, in contrast to general sparing of markers for neurons known to be localized to neurons intrinsic to the cerebral cortex and hippocampal formation.

An important role played by the cortical and hippocampal cholinergic deficits in the pathophysiology of AD has been further supported by behavioral and psychopharmacologic studies. Administration of the muscarinic acetylcholine receptor antagonist scopolamine to young healthy adults reproduces many aspects of the cognitive impairments seen in the aged. In particular, impairments of recent memory with relative preservation of verbal performance in cognitive tests are noted. In experimental animals, lesions of the basal forebrain cholinergic projections either to the hippocampal formation or to the cortex result in selective impairments in recent or working memory with preservation of long-term memory.

Etiology of AD

Approximately 15–20% of probands affected with AD have a clear family history for the disorder, compatible with an autosomal dominant pattern of inheritance. Recent studies suggest that the genetic predisposition for the disorder may be underestimated due to death of family members before they live through the risk period. Using restricted criteria for the clinical diagnosis, J. Breitner and M. Folstein have found a 50% risk for AD in siblings of probands with AD. The role of genetic factors is further strengthened by the observation that individuals with Down's syndrome, trisomy 21, invariably develop the histopathology and synaptic neurochemical deficits of AD if they live into their fourth decade.

Genetic predisposition to AD does not preclude environmental factors in the etiology. There is little evidence, at present, that aluminum is a toxic agent responsible for AD since brains of AD patients do not exhibit elevated levels of aluminum. Nevertheless, aluminum appears to be concentrated in neurons that exhibit neurofibrillary tangles. The possible role of infectious agents has received increasing interest with evidence that slow viruses or atypical infectious agents known as prions are responsible for the progressive neurodegenerative disorders, Jakob-Creutzfeldt dementia and kuru in humans, and scrapie in sheep. Notably, genetic factors determine the latency and expression of scrapie in experimental animals, a finding that underlines the potential interaction between hereditary predisposition and infectious agents in AD.

Treatment

At present, there is no treatment that forestalls the inexorable progression of AD. Effective management is based upon structuring the environment to compensate for the cognitive limitations of the affected individual. Neuroleptics can often be helpful in treating secondary psychotic symptoms that complicate the disorder. Current research indicates that drugs that potentiate central cholinergic neurotransmission, such as the acetylcholinesterase inhibtor physostigmine, may reduce memory deficits in AD. The general utility of such pharmacologic strategies in clinical management of AD remains to be determined.

Further reading

Katzman R, Terry RD, Bick KL, eds (1978): *Alzheimer's Disease: Senile Dementia and Related Disorders*. New York: Raven Press

Coyle JT, Price DL, DeLong MR (1983): Alzheimer's disease: A disorder of cortical cholinergic innervation. *Science* 219:1184–1190

Katzman RD, ed (1983): *Biological Aspects of Alzheimer's Disease*. Cold Spring Harbor Laboratory

Amnesia

Larry R. Squire

Amnesia refers to difficulty in learning new information or in remembering the past. These impairments can have a functional origin, but more commonly they result from neurological injury or disease. Functional amnesia is a psychiatric disorder. The amnesia in this case is presumed to result from conflict and repression, and its presentation varies from individual to individual. Typically, a patient is admitted to the hospital in a confused or frightened state. Memory for the past has been lost, especially personal, autobiographical memory. Often, the patient cannot produce his or her own name. Yet, the ability to learn new material is almost always intact. After the disorder clears, usually within a week, the lost memories return except for the events of the day or two just prior to hospital admission. Rarely, the disorder lasts longer, and sizable pieces of the past remain unavailable.

Characteristics

Brain injury or disease produces a different, more consistent pattern of memory impairment. The characteristics of amnesia are determined by the facts of normal brain function. Neurological disorders of memory impairment commonly occur in the context of more global dementing disorders, which affect memory together with language, attention, visuospatial disorders, and general intellectual capacity. Amnesia can also occur as a strikingly circumscribed impairment in the absence of other cognitive deficits. Patients with this disorder have intact intelligence test scores, intact language and social skills, and an intact memory for a great deal of information from the remote past, especially childhood. In addition, immediate memory is intact. That is, amnesic patients have a normal ability to repeat back a string of six or seven digits. In fact, they can hold a small amount of information in mind for several minutes, provided the material can be rehearsed. In much the same way, a person with normal memory functions might hold a telephone number in mind for a few minutes until it can be written down. For this reason, amnesic patients can carry on a conversation and often appear quite normal to casual observation. The difficulty arises when information must be recalled after a distraction-filled interval, or when recall is attempted directly after the presentation of an amount of information that exceeds immediate memory capacity (for example, 9 or 10 digits). In this situation, normal subjects will usually recall more items than amnesic patients. Furthermore, if the items are repeated a number of times, normal subjects quickly become proficient, but amnesic patients may never learn the whole list.

Retrograde amnesia

Amnesia affects not only the ability to learn new material, but also the ability to recall material learned prior to the onset of amnesia. In some cases of amnesia, this retrograde aspect of memory loss is temporally limited, affecting memories from the recent but not the remote past. In addition, within the affected time period, the amnesia can be temporally graded, so that very recent memories are most affected and less recent memories are less affected. This point has been established most clearly for the amnesias associated with head injury, electroconvulsive therapy, and certain well-studied patients with bilateral lesions involving the medial temporal lobe. In other cases of amnesia, for example, Korsakoff's syndrome, retrograde amnesia can be extensive and cover the majority of past life. The extensiveness and severity of this remote memory loss does not appear to be related to the severity of anterograde amnesia, which raises the possibility that it is caused by different lesions from those that cause the anterograde amnesia.

Etiology

Amnesia can occur for a number of reasons, e.g., after temporal lobe surgery, chronic alcohol abuse, encephalitis, head injury, electroconvulsive therapy, anoxia, and rupture and repair of an anterior communicating artery aneurysm. The memory disorder appears to depend on disruption of normal function in one of two areas of the brain: the medial aspects of the temporal lobe and the diencephalic midline. When the damage is bilateral, amnesia is said to be global. It affects verbal and nonverbal material, regardless of the sensory modality through which learning occurs. When damage is unilateral, the amnesic disorder is material-specific rather than global. In the case of left-sided damage, the deficit is for verbal material. Memory for nonverbal material (for example, for faces and spatial layouts) is affected in the case of right-sided damage.

Spared learning and memory abilities

Although amnesia can be severely disabling, the deficit does not affect all kinds of learning and memory. First, amnesia spares many kinds of skill learning. Motor skills, perceptuomotor skills, and certain cognitive skills can be acquired at a normal rate. In one study, amnesic patients learned to read mirror-reversed words at a normal rate and then retained the skill for months, despite the fact that when they had finished they could not remember the words that they had read and in some cases could not remember that they had ever practiced the mirror-reading skill on a previous occasion.

A second example of spared memory ability is priming—the facilitation of performance by recent exposure to stimulus material. To illustrate priming, suppose that normal subjects are shown a short word list, which includes the words *motel* and *absent*. When they are then shown the fragments *mot--*

and *abs*---, with instructions to form the first word that comes to mind, they have a strong tendency (about 50%) to produce the words that were recently presented. (The probability is only about 10% that subjects will produce these words if they were not presented recently.) Amnesic patients exhibit this effect to the same extent as normal subjects. Moreover, in both amnesic patients and normal subjects, this effect declines gradually after word presentation and disappears in a few hours. Priming effects in both amnesic patients and normal subjects are modality bound, in that the effects are strongest when words are presented in the same sensory modality as word fragments, and significantly diminished (although not eliminated) when words and fragments are in different modalities.

The preservation of skills and priming in amnesia shows that these forms of learning and memory do not depend on the medial temporal and diencephalic structures damaged in amnesia. These findings have raised the possibility that multiple forms of memory exist. The kind of memory that is affected in amnesia is explicit and accessible as conscious memory. For this reason, it has sometimes been termed declarative memory. It can be declared, or brought to mind, as a proposition or as an image. In declarative learning, new facts or events are added to memory, or associations are formed between already available but unrelated items. The kind of memory that is preserved in amnesia has been termed procedural. It is implicit, expressible only in performance. The knowledge is embedded in the procedures that are engaged when learning occurs. In priming, a preexisting representation is temporarily facilitated. Newly formed representations can also be primed in some cases. In skill learning, preexisting procedures are combined to form long-lasting cognitive operations.

Animal models

Animal models of human amnesia have been established in the monkey. They could probably be achieved in any mammal and possibly in other vertebrates, such as birds. Monkeys with bilateral medial temporal lesions exhibit a profound impairment on the kinds of tasks of new learning ability that human amnesic patients fail. The same animals succeed at tasks of motor skill learning. They also do well at learning pattern discriminations, which share with motor skills the factors of incremental learning and repetition over many trials. Monkeys with bilateral medial thalamic lesions also exhibit an amnesic disorder. Lesions of the mammillary nuclei produce a small, negligible impairment. Damage to other regions such as ventromedial frontal cortex and basal forebrain, which have strong anatomical links to the medial temporal region and medial thalamus, also appears to produce memory impairment. Further work is needed to quantify each kind of impairment, in order to compare the effects of each lesion and to determine what qualitative differences might exist in the memory impairment associated with each lesion. In addition, further work is needed to determine which lesions are required to produce a stable memory impairment that can last for years.

Hippocampus

Occasionally, useful information about the neuroanatomy of amnesia comes from well-studied cases of human amnesia, where extensive neuropsychological and neuropathological data are available. Case R.B. became amnesic in 1978 following an ischemic episode that occurred as a complication of cardiac by-pass surgery. When he died in 1983, thorough analysis of his brain revealed a bilateral lesion confined to the CA1 field of hippocampus. The lesion extended the full rostral-caudal length of the hippocampus, involving an estimated 4.63 million pyramidal cells. There was no other clinically significant lesion. This case shows that a lesion limited to the hippocampus is sufficient to cause amnesia. This level of impairment can be exacerbated by additional damage. For example, the well-known surgical patient H.M., who appears to be as severely amnesic as any other thoroughly studied case, has bilateral damage to parahippocampal gyrus, entorhinal cortex, and amygdala, in addition to hippocampal damage.

Amnesia as a window on normal memory functions

The analysis of memory disorders in neurological patients, and in animal models of amnesia, continues to provide useful information about how the brain accomplishes learning and memory. Amnesia appears to result from damage to a brain system that performs a particular function, or computation, when learning occurs. This neural system must work in concert with the distributed sites in neocortex and elsewhere that are believed to be the loci of memory storage. When sufficient time has passed after learning, the distributed sites that constitute memory for an event are able to support an enduring representation in memory on their own, without the participation of this neural system. Retrieval can then be accomplished in the absence of the neural system damaged in amnesia. Better understanding of this neural system, and its component parts and connections, should also illuminate neurobiological studies that are conducted at more elemental levels of analysis by establishing a clearer connection between neurobiological data and functional questions about the organization of memory in the brain.

Further reading

Mishkin M (1982): A memory system in the monkey. *Phil Trans R Soc Lond B*, 298:85–95

Schacter DL (1985): Multiple forms of memory in humans and animals. In: *Memory Systems of the Brain*. Weinberger N, McGaugh J, Lynch G, eds. New York: Guilford

Squire LR (1987) *Memory and Brain*. New York: Oxford

Zola-Morgan S, Squire LR (1985): Complementary approaches to the study of memory: human amnesia and animal models. In: *Memory Systems of the Brain*, Weinberger N, McGaugh J, Lynch G, eds: New York: Guilford

Amphetamines

Norman J. Uretsky

The amphetamines are a group of central nervous system (CNS) stimulants, which produce most of their biological effects by enhancing neurotransmission at catecholaminergic synapses in the peripheral and central nervous systems. The term amphetamines usually includes dextroamphetamine, levoamphetamine, racemic amphetamine, and methamphetamine. Amphetamine (α-methylphenethylamine) is the prototype drug of this class and exists in 2 stereoisomeric forms. The dextroisomer, d-amphetamine, is about 3 times more potent as a CNS stimulant than the levo form, l-amphetamine. In contrast, the latter compound has more potent cardiovascular effects. Methamphetamine (N-methyl α-methylphenethylamine) is a more potent CNS stimulant than d-amphetamine. There are several other chemical analogs of amphetamine that produce similar biological effects. These include methylphenidate, pipradrol, and a variety of compounds that are presently marketed for their appetite suppressant effects.

By enhancing noradrenergic neurotransmission, amphetamine causes the stimulation of organs innervated by postganglionic sympathetic neurons, resulting in a variety of effects that mimic the activation of the sympathetic nervous system. Since the therapeutic effects of amphetamine and its structural analogs are related to their actions in the central nervous system, these autonomic effects would tend to limit the therapeutic usefulness of these drugs.

Amphetamines, which are lipid soluble in the free-base form, readily cross the blood brain barrier and produce a variety of behavioral changes in humans and animals. A characteristic feature of their CNS stimulant action is an arousal or alerting response, which may account for their ability to reduce fatigue and to enhance physical and mental performance. Amphetamines also reduce rapid eye movement sleep, which together with the arousal response may explain their effectiveness in the treatment of narcolepsy. Amphetamine stimulates locomotor activity, an effect which appears to be mediated in rodents through the enhancement of dopaminergic neurotransmission in the nucleus accumbens. At higher doses, amphetamine elicits a compulsive stereotyped behavioral response which has been primarily related to an enhancement of dopaminergic neurotransmission in the corpus striatum. The behavioral characteristics of this response depend on the species being studied. For example, rats injected with high doses of amphetamine display repetitive sniffing, licking, and gnawing behavior, while humans engage in compulsive repetitive manipulatory tasks that appear to serve no useful purpose.

Another characteristic effect of amphetamines, which underlies their abuse potential, is an elevation of mood or euphoria, which is associated with an enhanced feeling of energy and self-confidence. After intravenous administration, this effect has been described as extremely intense and pleasurable. The euphoria produced by amphetamine has been related to the enhancement of noradrenergic and dopaminergic neurotransmission at synapses derived from neurons of the medial forebrain bundle. Since animals will lever press in order to electrically stimulate the medial forebrain bundle, it has been postulated that neurons of this tract are part of a reward system in the brain, which functions in the regulation of mood and motivation.

Amphetamines can suppress appetite. This anorexic action appears in part to be due to a direct action on neurons on the lateral hypothalamus or feeding center. A variety of more "selective" anorexic phenethylamine analogs of amphetamine have been developed that possess relatively less central stimulant properties. Recently, a high-affinity, saturable, and stereospecific binding site for d-amphetamine has been demonstrated in membrane preparations from rodent hypothalamus. The potencies of the various structural analogs of amphetamine in inhibiting ^{3}H-d-amphetamine binding correlate well with their potencies in producing anorexia, suggesting that these binding sites mediate the appetite suppressant effects. Clinically, however, these analogs still exhibit the same side effects as amphetamines, and tolerance develops rapidly to their anorexic effects. Therefore these drugs are unlikely to be useful in a long-term weight reduction program.

Although amphetamine is a central stimulant, it has been found to be useful in the treatment of hyperactive children diagnosed as having attention deficit disorders. In this condition, the paradoxical "calming" effect produced by amphetamines may be a result of an improvement in their ability to concentrate. A paradoxical beneficial effect of amphetamine has also been demonstrated in petit mal epilepsy.

High doses of amphetamines or repeated administration of these drugs produce a reversible toxic psychosis, characterized by paranoid behavior that closely resembles schizophrenia. There may be auditory and tactile hallucinations; the latter has been described as the sensation of insects or snakes crawling underneath the skin. Since evidence has been accumulating that antipsychotic drugs produce their therapeutic effects by blocking dopaminergic receptors, and since amphetamine enhances dopaminergic neurotransmission, it is likely that the paranoid psychosis induced by amphetamines is mediated through their ability to enhance dopaminergic neurotransmission.

Recent studies have shown that the repeated administration of high doses of amphetamine can produce neurotoxic effects in the brains of several animal species. Thus, amphetamines in high doses produce a long-lasting reduction in brain tyrosine hydroxylase and tryptophan hydroxylase activities and in catecholamine transport, suggesting damage to catecholamine- and serotonin-containing neurons. These biochemical changes have been supported by morphological evidence of nerve degeneration. At present, the behavioral consequences of these neurotoxic effects of amphetamines are unclear.

Tolerance develops to almost all the effects of amphetamines except the toxic psychosis which becomes progressively more intense with continuous drug administration or the intake of high doses. After abrupt withdrawal following the chronic intake of amphetamine, the user will experience an intense fatigue, depression, feelings of hunger, and a strong desire to continue taking the drug. At present, it is not clear whether these symptoms constitute an abstinence syndrome.

Almost all the effects of amphetamines are thought to be due to their actions on noradrenergic and dopaminergic neurons, although at high doses actions on serotonergic neurons may also play a role. Amphetamine, which structurally resembles norepinephrine and dopamine, can cause the release of these catecholamines from presynaptic nerve terminals, thus increasing their concentration in the synaptic cleft. In addition, amphetamine inhibits the specific carrier-mediated uptake process for these amines. Since reuptake of catecholamines into the presynaptic nerve terminal appears to be the major mechanism for terminating their synaptic action, the inhibition of reuptake of the released catecholamines by amphetamines should further increase their concentration and activity in the synaptic cleft. Based on studies with reserpine, which depletes dopaminergic storage sites (but does not block the behavioral effects of amphetamine), and α-methyl-p-tyrosine, which inhibits dopamine synthesis (and blocks the behavioral effects of amphetamine), the dopamine released by amphetamine plays a major role in its CNS actions, and appears to be mainly derived from a small reserpine-insensitive cytoplasmic pool which is continually replenished by new synthesis. Amphetamine, which can stimulate dopamine synthesis, may prevent the newly synthesized dopamine pool from being depleted as a result of the enhanced dopamine release. The precise mechanism behind the amphetamine-induced release of dopamine and norepinephrine is still unclear. Recent studies have suggested that amphetamine, by binding to the bidirectional catecholamine uptake carrier, might increase the concentration of this carrier on the inside of the membrane, thereby accelerating the carrier-mediated efflux of catecholamines.

In contrast to the amphetamines, the CNS stimulant effects of some of its structural analogs (for example, methylphendiate) are markedly antagonized by the prior administration of reserpine but are relatively unaffected by α-methyl-p-tyrosine. Since the CNS stimulant effects of both the amphetamines and its analogs are antagonized by dopamine receptor blocking agents, it is thought that the structural analogs may act differently on dopaminergic neurons to enhance dopaminergic neurotransmission. One explanation is that the amphetamine analogs act mainly by blocking the reuptake of dopamine released from a granular (reserpine-depletable) pool and have a much weaker direct releasing action from the reserpine-insensitive cytoplasmic dopamine pool.

Further reading

Ridley RM (1983): Psychostimulants. In: *Psychopharmacology*. Part I: *Preclinical Psychopharmacology*, Grahame-Smith DG, Cowan PJ, eds. Amsterdam, Oxford, Princeton: Excerpta Medica

Iversen LL, Iversen SD, Snyder SH, eds. (1978): *Handbook of Psychopharmacology (Stimulants) 11*, New York: Plenum Press

Kalant OJ (1973): *The Amphetamines: Toxicity and Addiction*, 2nd ed, Toronto, Buffalo: University of Toronto Press

McMillen BA (1983): CNS stimulants: two distinct mechanisms of action for amphetamine-like drugs. *Trends Pharmacol Sci* 4:429–432

Antidepressants

Elliott Richelson

Antidepressants are drugs which are effective for treating depression, a serious psychiatric disorder that afflicts about 5% of the adult population in the United States. These drugs, which are also less commonly called psychoanaleptics, thymoleptics, or mood-elevating drugs, are represented by a diverse group of chemical structures (see Fig. 1) whose mechanism of therapeutic action is currently unknown. However, there are many hypotheses about how they work, and in general, these theories involve biogenic amine neurotransmitters, especially norepinephrine and serotonin.

Of the many types of antidepressants, one of the first discovered and the most widely used types is the tricyclics type, so named because of the three fused rings in their structures. This class has been further subdivided according to the type of amine on the side chain of these compounds into tertiary amines (for example, imipramine and amitriptyline) and secondary amines (for example, desipramine and protriptyline). The prototype of this class of compounds, imipramine (Fig. 1), was initially synthesized for use as an antihistamine (H_1) but was discovered to have antidepressant properties as the result of astute clinical observations on depressed schizophrenic patients. After chlorpromazine was shown to be effective as an antipsychotic drug in the early 1950s, researchers sought other drugs to treat psychosis and selected imipramine for a trial because of its close structural similarity to chlorpromazine. Though imipramine was ineffective as an antipsychotic drug, it made depressed schizophrenic patients less depressed. This finding led to clinical trials with imipramine which proved its efficacy as an antidepressant.

Another class of antidepressants, the monoamine oxidase inhibitors, was introduced also in the 1950s as a result of observations in both the clinic and the laboratory. The first compound of this group to be tested and proved effective as an antidepressant was iproniazid, an antituberculous drug, after it was observed that it caused euphoria and elation in some patients treated for tuberculosis and that it reversed the apparent sedation caused by reserpine in laboratory animals. This result with the reserpine-treated animal (usually mouse or rat) eventually led to its use as a model for the depressed human (perhaps in part because of the psychomotor retardation seen in many depressed patients). Thus, the reversal of reserpine's effects by drugs became a screening test for new antidepressants. Some justification for the use of this model is found in the fact that reserpine causes depression in a small percentage (about 10%) of patients who are being treated with this drug for hypertension. In general, however, biological research in psychiatry is hindered by the lack of good models of human mental disease.

Monoamine oxidase inhibitors are much less frequently used than other types of antidepressants because of the "cheese reaction," a marked and potentially fatal increase in blood pressure that can occur when a patient being treated with one of these drugs ingests foods high in tyramine (such as aged cheese). The excess tyramine, unable to be degraded by monoamine oxidase, displaces neuronal stores of catecholamines which ultimately increase blood pressure. Consequently, patients taking monoamine oxidase inhibitors are required to be on a tyramine-free diet and must also avoid the use of certain sympathomimetic stimulants, which are often found in over-the-counter cold remedies.

Sympathomimetic stimulants such as amphetamine and methylphenidate are not considered antidepressants although they are used by some clinicians in special cases. In addition, lithium ion, administered as a salt, has not been established as an effective antidepressant, although it has unquestioned efficacy in the treatment of mania and the prevention of manic-depressive illness. Likewise, amino acids such as L-tryptophan and L-dopa have not been generally proved to have antidepressant effects.

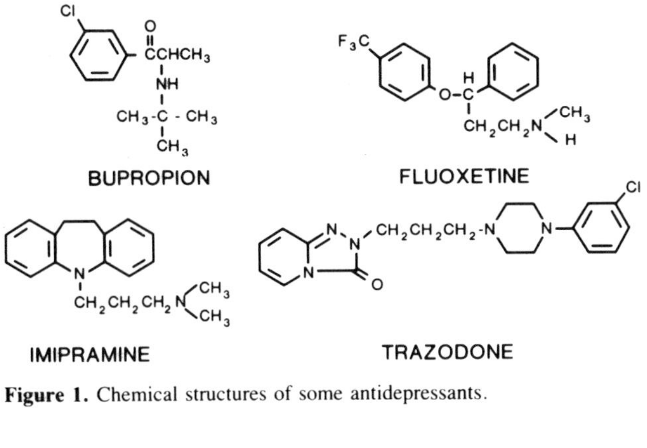

Figure 1. Chemical structures of some antidepressants.

Classification of antidepressants

In earlier years antidepressants were classified as either tricyclics or monoamine oxidase inhibitors. This method of classification, however, mixes structural and functional criteria, and ideally it would be better to classify these drugs on the basis of either structure or function. Unfortunately, the newer, so-called second-generation antidepressants (such as bupropion, fluoxetine, and trazodone) (Fig. 1) have diverse structures and no apparent common neuropharmacologic effect, so that it is difficult to arrive at a satisfying classification that encompasses all antidepressants. One temporary approach is to classify these drugs on the basis of the total number of rings in the structure. For this purpose such terms as bicyclic, tricyclic, or tetracyclic can be used, in order to describe a structure with 2, 3, or 4 fused rings. The term heterocyclic, which describes a compound with a ring containing at least one atom different from carbon, is not useful since to classify the antidepressants as heterocyclic and nonheterocyclic subdivides the tricyclic type (for example, imipramine is heterocyclic but amitriptyline is not).

Pharmacologic effects of antidepressants: Focus on the synapse

It is well established that antidepressants affect the synaptic functions of neurotransmitters by various mechanisms. Neurotransmitters are small molecules that neurons use to communicate with one another; synapses are structures by which nerve terminals abut onto other neurons. Neurotransmitters are usually amino acids or their derivatives and are released from the nerve terminal to diffuse into the synaptic cleft to interact with specific receptors on the outside surface of the target cells. These receptors can be located both postsynaptically and presynaptically (autoreceptors). In addition, some of the released neurotransmitter is recovered by the nerve ending (a process called uptake or reuptake), thereby preventing the neurotransmitter from overstimulating a receptor. Rapid degradation of the neurotransmitter can also reduce the level of the molecule in the synaptic cleft.

Not long after the discovery of tricyclic antidepressants, researchers demonstrated that these drugs acutely blocked uptake by the nerve ending of both norepinephrine and serotonin. It was considered, then, that the therapeutic effect of the antidepressant was due to increased levels of released neurotransmitter in the synaptic cleft. The inhibition of uptake by tricyclic antidepressants is one of the cornerstones of the so-called biogenic amine hypothesis of affective illness, which (in the simplest terms) states that a deficiency of certain biogenic amines (for example, norepinephrine) at functionally important synapses causes depression, while an excess causes mania. Other data in support of this hypothesis are that monoamine oxidase inhibitors increase brain levels of catecholamines and serotonin by preventing their degradation; whereas reserpine, which as noted above can cause depression in some patients on this drug for hypertension, depletes brain levels of these molecules.

Earlier data indicated that tertiary amine tricyclic antidepressants were potent blockers of serotonin uptake but weak blockers of norepinephrine uptake, whereas the converse appeared to be true for the secondary amine compounds. These data led to the hypothesis that there are serotonin- and norepinephrine-deficient depressions. Despite decades of research, clinical data to support this theory have not been conclusive. In addition, more recent data on uptake blockade by tricyclic and other antidepressants in vitro show that most of these compounds (including tertiary amine tricyclic antidepressants) are more potent at blocking uptake of norepinephrine than at blocking uptake of serotonin, and some of the newer antidepressants are very weak at uptake blockade. These data do not support the hypothesis subdividing antidepressant action at serotoninergic and noradrenergic systems.

More recent research on other acute pharmacologic effects of antidepressants has concentrated on their ability to block receptors. Drugs that block receptors (antagonists) cause their effects by preventing access of the neurotransmitter to its receptor. How tightly a drug binds to a receptor (the "affinity") can be determined by a variety of techniques. The higher the affinity, the greater the effect of the drug, either desirable or undesirable. Antidepressants are antagonists of several different types of receptors including histamine (H_1 and H_2), muscarinic acetylcholine, α_1- and α_2-adrenergic, dopamine D_2, and serotoninergic (S_1 and S_2) receptors. In general, their most potent effect is at the histamine H_1 receptor, and some antidepressants (especially the tricyclic antidepressant doxepin) have such high affinity for this receptor that they are among the most potent histamine H_1 antagonists available. This finding has led to their use outside psychiatry for the treatment of allergic and certain dermatological problems.

Acute receptor antagonism by antidepressants is probably not related to the mood elevating effects of these drugs but underlies various adverse effects. For example, muscarinic receptor blockade can cause dry mouth and constipation; and histamine H_1 receptor antagonism can cause sedation and drowsiness. However, for some depressed patients who are also agitated, sedation is a desirable effect of the antidepressant. These drugs vary greatly in their potential to cause these and other adverse effects, and it is possible to predict the likelihood that a particular antidepressant will cause a specific side effect from the knowledge of its affinity (from in vitro studies) for the particular receptor involved. Since the second-generation drugs in general tend to have lower affinities for receptors, they are less likely to cause the adverse effects commonly seen with the older antidepressants.

The mood-elevating effect of antidepressants does not become apparent until at least 10 days to several weeks after treatment has started, whereas the pharmacologic effects discussed above are apparent within hours after administration of drug. Thus, researchers have sought to find biological actions of antidepressants which are evident only after chronic drug administration. One such long-term effect is the decreased sensitivity (desensitization) of certain brain neurotransmitter receptors which in most cases leads to a reduction (down-regulation) in numbers of receptors. Down-regulation of receptors in rat brain has been demonstrated to occur with chronic administration of many different types of antidepressants as well as with chronic electroshock. The receptors most consistently affected are those for the catecholamines. Serotonin receptors may also be reduced with chronic antidepressant administration to rats. In addition, serotoninergic innervation may play a permissive role in catecholamine receptor desensitization, since the reduced sensitivity of catecholaminergic receptors caused by antidepressants does not occur when serotonin-containing neurons are destroyed.

Problems with the desensitization hypothesis of antidepressant action exist because some of the newer antidepressants such as bupropion, fluoxetine, and trazodone (Fig. 1) do not appear to cause desensitization. In addition, the antipsychotic chlorpromazine causes desensitization but has little or no efficacy as an antidepressant.

It is clear that we must focus on synaptic function in order to discern the mechanisms underlying the therapeutic effect of antidepressants. However, since the particular neurotransmitter system involved in affective behavior remains undiscovered, further research is required before the sites of therapeutic action of antidepressants are known and their mechanisms of action understood.

Further reading

Enna SJ, Malick JB, Richelson E, eds (1981): *Antidepressants: Neurochemicals, Behavioral and Clinical Perspectives.* New York: Raven Press

Richelson E (1982): Pharmacology of antidepressants in use in the United States. *J Clin Psychiatry* 43(11) Sec 2:4

Anxiety and Anxiety Disorders

David V. Sheehan and Kathy H. Sheehan

Definition

Phenomenologically, anxiety may refer to an emotion, a feeling, a symptom, or a cluster of cognitive and somatic symptoms. Etiologically, it may describe reactions to danger, stress, or conflict, the results of trauma or frightening memories, the toxic withdrawal reactions to many drugs and illnesses, a habit (a persistent pattern of maladaptive behavior acquired by learning), or the symptomatic expression of a genetically inherited metabolic disease. The prevailing opinion in psychiatry views *anxiety* as a cluster of symptoms that impairs normal functioning.

Classification

Many classification systems of anxiety disorders have been suggested. The most widely used classification of anxiety disorders is that of the *Diagnostic and Statistical Manual of Mental Disorders*, third edition (DSM III), of the American Psychiatric Association, published in 1980. For practical purposes, four types of anxiety are identified in most anxiety classifications:

1. An adjustment or response to immediate stress/danger or conflict anxiety (adjustment disorder with anxious mood).
2. An isolated anxiety that is bound to a stimulus, is out of proportion to the reality of the situation, and leads to stimulus avoidance (phobic anxiety/simple phobia).
3. An anxiety that is primarily autonomous—unattached to any apparent stimulus at least some of the time (panic disorder/endogenous anxiety).
4. Anxiety that is secondary to other medical or psychiatric disorders or drugs.

Whether each of these groups is homogeneous or is made up of several subgroups is a matter of debate.

Clinical description

Response to stress or conflict anxiety is characterized by symptoms in direct and immediate relation to the stress. The symptoms include dry mouth, increased heart rate and perspiration, tremor of hands, fluttery stomach, increased muscle tension, and cognitive fear/anxiety. When the stress is removed the symptoms usually remit.

In conditioned or simple phobic anxiety the symptoms are similar, but there is crescendo anxiety on exposure to a stimulus and it is out of proportion to the reality of the situation and continues to lead to avoidance behavior. Often there is generalization of the anxiety from the core stimulus.

The pathological anxiety that is associated with unexpected/unprovoked anxiety attacks (panic disorder) and multiple phobias is associated with attacks of symptoms that are more profoundly disruptive and disabling to the patient. These include sudden bursts of tachycardia, lightheadedness, dizzy spells, imbalance, difficulty getting a breath, chest pain or discomfort, choking or smothering sensation, hot flashes, nausea, or a feeling that the surroundings are strange, unreal, detached, or unfamiliar. There is frequently a severe mental panic with a sense of loss of control or imminent doom and a flight response. While this is occurring the victim cannot identify any immediate obvious justifiable precipitant for the attack. The suddenness of the surge of symptoms and their occurrence without warning or clear-cut precipitation sets this disorder apart from response to stress anxiety.

Typically, the natural history of panic disorder progresses through several stages. It often begins with sudden attacks of one or two symptoms without cognitive anxiety. It then progresses to sudden attacks with many symptoms and a feeling of mental panic. At a loss to identify psychological precipitants, patients may believe they have a medical disease and cannot be reassured. Next the severe attacks become associated with specific situations (where the severe attacks occurred), and limited phobic avoidance to these situations starts. As the attacks persist there is progression to extensive phobic avoidance (including social phobias and agoraphobia) and increasing depression.

Epidemiology

Anxiety and phobic disorders are the most prevalent of all psychiatric disorders. A large multicentered epidemiological catchment area study, using precise and conservative diagnostic criteria, found the lifetime prevalence of anxiety and phobic disorders at 8.3% of the population over 18 years. Seven percent of the adult population over 18 suffer from a phobic disorder alone. With the current adult population of the United States at 240 million, it is estimated that 19.9 million have suffered from an anxiety/phobic disorder and 16.8 million from a phobic disorder during their lifetime. Among women it ranked as the number one psychiatric disorder in all age groups.

Within a six-month period, 77% of those with an anxiety and phobic disorder visit a health or mental health care facility to seek medical or psychiatric evaluations. Another National Institute of Mental Health study found that 11% of the general population are using antianxiety medications and over 70 million prescriptions for tranquilizers are written annually.

Response to stress anxiety and simple phobias may start at any age. In contrast, the anxiety disorder associated with unexpected anxiety attacks (panic disorder) and multiple phobias (e.g., agoraphobia) has an unusual age of onset with a peak age of onset in the 20s. It rarely begins before the age of 14 or after 40.

Response to stress anxiety and simple phobias is more evenly distributed between men and women—with just over 60% being women. Among those with unexpected anxiety attacks and multiple phobias 75–80% are women; 73–92% are symptomatic when reevaluated up to 20 years after the initial diagnosis.

Etiology

Simple phobias and response to stress without unexpected anxiety attacks are believed to be acquired traumatically or by conditioning. This can happen to any normal person suitably exposed.

Multiple phobias with unexpected anxiety attacks are believed to result from a metabolic disease to which there is a genetic vulnerability. In this class of phobias there appears to be an interaction of three causative forces: (1) the biological core; (2) conditioning that arises as a complication of this biological core; and (3) environmental stress. In any one individual each of these may operate to varying degrees. The prevailing view is that the biological force is the largest contributor and that environmental stress may even be absent.

Evidence from several family studies shows that approximately 15–41% of first-degree relatives of affected individuals have the condition. When both parents are affected, the pooled prevalence is 40%. There is an increased concordance in identical, as compared to nonidentical, twins. This suggests that genetic factors override environmental influences. Several biological models of the anxiety disorder associated with unexpected/unprovoked anxiety attacks have been suggested. There are three theories (not necessarily mutually exclusive) about the biological core of this disorder that show promise. Each is backed by a body of scientific data and each probably represents a different part of the total and still incomplete picture.

The first biological model, called the locus coeruleus model, posits that anxiety attacks arise due to arousal of the locus coeruleus. This small nucleus in the pons of vertebrate brain stem contains at least 50% of all neurons in the central nervous system using norepinephrine as a neurotransmitter. It has numerous projections to many areas of the brain. Electrical or drug (yohimbine) stimulation of the locus coeruleus tends to produce anxiety attacks and a fear response. Anxiety and fear-like responses are blocked by locus coeruleus lesions or ablation.

The second biological model is the gamma-aminobutyric acid (GABA)-benzodiazepine model. It suggests that the *GABA*-benzodiazepine receptor is malfunctioning in pathological anxiety. This may be due to a deficiency in the quality or quantity of the inhibitory neurotransmitter, in the receptor sensitivity to the neurotransmitter, or in the quality or quantity of the receptors or their adjacent ion channels. Benzodiazepines, which calm anxiety, and beta carbolines and caffeine, which precipitate anxiety, are believed to act at this site.

The redox hypothesis suggests that panic disorder may result from a pathological oversensitivity of central chemoreceptors in the ventral medullary center to CO_2 or lactate or from a lability in the first steps of the ventilation arousal cascade which those chemoreceptors activate. Lactic acid is the end product of the anaerobic metabolism of carbohydrates and is derived solely by the reduction of pyruvate to lactate. The process requires oxidation of the CO factor NADH (nicotinamide adenine dinucleotide, reduced). The NADH/NAD ratio is the redox state of the site where the reaction is taking place, usually within the mitochondria. Rises in the lactate/pyruvate ratio with concomitant falls in chemoreceptor intraneuronal pH is hypothesized as the trigger. Experimentally, lactate infusion or breathing 5–7% CO_2 may lead to a rapid fall in pH at the ventral medullary center with the production of panic attacks in susceptible patients.

The evidence directly identifying a physical defect in patients with unexpected anxiety attacks is still at an early stage of investigation. One recent study gave an 0.5 M solution of sodium lactate to victims of panic disorder and normal controls. Most of the panic patients had anxiety attacks after approximately 12 minutes of the infusion, but normal persons did not. During their panic attacks, there was a significant right-left shift in the blood flow in one small area of the brain—the parahippocampal region—on positron emission tomography (PET) scan. This did not occur in normal controls. For the first time this showed that panic attacks in these patients could be seen with this imaging technology. Although the central defect appears to be a neurochemical one it has not yet been precisely identified.

As a result of having unexpected, unprovoked attacks of anxiety, patients begin to fear and avoid the situations they associate with their severe attacks. That is, these biologically based attacks condition them to become phobic. The more intense the attacks and the longer they continue, the more phobias the patient acquires. Eventually many patients become housebound and disabled by the extensive phobic avoidance.

Stress appears to be a nonspecific aggravator of almost all medical illnesses. Stress will magnify panic disorder and extensive phobic avoidance when it is present. However, the view that all those with extreme phobias and pathological anxiety have some stress or conflict in their lives seems incorrect. Some do. Frequently however, their stresses do not differ substantially from anyone else's.

Complications

Until recently, it was widely believed that anxiety disorders were not associated with any medical complications beyond the immediate subjective discomfort of the symptoms. However, recent evidence suggests that panic disorder has an excess mortality when compared to matched normal controls. This excess mortality comes from two sources—suicide and, particularly in men, cardiovascular death. Although those with panic disorder are generally not seen as depressed, over a lifetime they appear to be as likely to kill themselves as those with depressive illness. They are at greater risk to develop hypertension. Approximately 34% have an associated mitral valve prolapse, and approximately 20% have a history of alcohol or substance abuse—usually in an attempt to control their symptoms.

Treatment

Stress or conflict anxiety is a response to justifiable life events. The treatment of choice is psychotherapy that is supportive and oriented to resolution of the conflict.

Simple phobias without unexpected anxiety attacks are conditioned rather than biologically based. The treatment of choice is behavior therapy, particularly exposure. Medications are rarely necessary.

Multiple phobias with expected anxiety attacks usually require a combination of several treatments. About 15% of cases have a spontaneous remission—that is, the symptoms of the disorder clear up for no particular reason. These 15% often ascribe their cures to whatever positive treatment they are engaged in at the moment. However, the majority of the victims of this illness, if they are to get full relief of symptoms, require an antipanic medicine, often additional exposure to behavior therapy, and sometimes psychotherapy. There are four major steps in treatment of panic anxiety with phobias.

Step 1. Control the metabolic core of the disorder with one of the following classes of antipanic medications: (a) tricyclic antidepressants, (b) monoamine oxidase (MAO) inhibitors, (c) alprazolam and clonazepam, or (d) trazodone. Each of these has advantages and disadvantages. Most have disruptive side effects. They require considerable skill to use correctly and to full advantage. They should block the core unexpected anxiety attacks and help lessen the phobic anxiety.

Step 2. Extinguish any remaining phobic avoidance with exposure behavior therapy. Sometimes patients can do this on their own without the assistance of a behavioral psychologist.

Step 3. If psychological or social problems complicate recovery, then additional psychotherapy may be necessary to help resolve remaining difficulties.

Step 4. Because this is a long-term relapsing disorder in the majority of cases, these patients require long-term monitoring to ensure that they do not relapse and become disabled again. Because they have often acquired much incorrect information about their condition, it is important to correct this with proper patient education about the disorder and its treatment. Correct information is one of the best insurances against future relapse and disability.

Further reading

Tuma AH, Maser JD, eds (1985): *Anxiety and Anxiety Disorders*. New Jersey: Lawrence Erlbaum Associates

Sheehan DV (1983): *The Anxiety Disease*. New York: C Scribners Sons; rev paperback ed Bantam Books, 1986.

Klein DF, Rabuin JG eds (1981): *Anxiety: New Research and Changing Concepts*. New York: Raven Press

Autism

Edward M. Ornitz

Autism is a severe pervasive developmental disorder of behavior which is not accompanied by demonstrable neurological signs, consistent neuropathology, metabolic disorder, or genetic markers. Four in 10,000 children are afflicted and about 80% of those afflicted are mentally retarded. The onset is usually within the first 30 months of life. Most patients remain severely disabled and require custodial care throughout their lives. There is no specific treatment; management is primarily behavior modification carried out in special educational or residential programs, and through parent training. Longevity is within the normal range.

The behavioral syndrome is unique, consisting of specific disturbances of social relating and communication, language, and cognition, response to objects, sensory modulation, and motility. The complete syndrome is usually observed, or described retrospectively by the parents, before 5 years of age. During later childhood and adolescence, the clinical picture may change, such that some patients continue to appear primarily autistic and others more retarded; a higher functioning minority develop schizoid personality disorders. Although originally described as the earliest manifestation of schizophrenia, most authorities consider autism and schizophrenia to be phenomenologically as well as genetically distinct syndromes. However, recent developmental studies show that some high-functioning verbal autistics develop a characteristic schizophrenic thought disorder. Behavioral analyses point to common elements in the responses to sensory input of young autistic children and older schizophrenics, suggesting a common dysfunction of sensory modulation and subsequent information processing. The disturbances of relating to people and objects include emotional remoteness, lack of eye contact, indifference to being held, stereotypic ordering and arranging of toys without regard to their function, intolerance of change in surroundings and routines, and the absence of imaginative play. The disturbances of communication and language include the absence of both verbal and nonverbal communicative intent, severe delays in the acquisition of language, and deviant forms of language, such as delayed echolalia and pronoun reversal. The disturbance of sensory modulation involves all sensory modalities, and the faulty modulation is manifest as both under- and overreactivity to sensory stimuli. The latter is often associated with a tendency to seek out and induce sensory input, e.g., visual scrutiny of spinning objects. Some of the motility disturbances, e.g., hand flapping, may provide such input through proprioceptive and kinesthetic channels. The disturbances of sensory modulation and motility occur predominantly, though not exclusively, in 2- to 4-year-old autistic children. Before the age of 5 years, the disturbances of sensory modulation and motility are observed with almost the same frequencies as the disturbances of relating to people and objects.

The uniqueness of this syndrome suggests one underlying pathophysiological mechanism. However, multiple etiologies, which could activate or replicate such a mechanism, are suggested by the association with autism of many prenatal, perinatal, and neonatal conditions which putatively are likely to insult fetal or neonatal brain function. In particular, congenital rubella, toxemia, neonatal anoxia, and infantile spasms have been implicated. Such conditions account for about one-quarter of all cases. In the remaining cases, potential etiologic factors have not been identified, although there is some evidence from family studies suggesting a subgroup with a genetic component. There is a paucity of pathologic studies, and the few autistic brains that have been examined have failed to reveal consistent neuropathology. Biochemical investigations have not revealed any consistent neuromodulator or neurotransmitter abnormalities, although elevated blood serotonin in about 25% of cases remains an unexplained finding.

The symptomatology of autism, particularly the disturbance of sensory modulation, and some neurophysiological studies suggest that this syndrome is, in the absence of consistent neuropathology, a true neurophysiological disease, and that the hypothesized pathoneurophysiological mechanism is at the interface between sensory processing and information processing. Neurophysiological and clinical studies of this mechanism have taken two general directions, one stressing the disturbances of language and cognition and the other stressing the disturbances of sensory modulation and motility.

Neurophysiological studies of cortical events are relevant to the autistic disturbances of language and communication, and to an underlying postulated specific cognitive disorder, presumably of cortical origin. This research has included electroencephalographic (EEG), radiological, including computerized axial tomography (CAT), and event-related potential studies. Both hemispheric lateralization and nonlateralizing phenomena have been studied. The possibility of dysfunction of hemispheric lateralization has been considered because of the language disorder in autism. The severe delays and deviances in language development suggest pathophysiological cortical mechanisms, although the mesencephalic and tectal portions of the reticular core within the brain stem have been implicated in the autistic failure to perceive prosodic speech features which are essential for social communication. Generally, inadequate modulation of sensory input during subcortical processing would compromise its value as information during cortical processing. Although this hypothesis suggests that abnormal cortical processing is a secondary consequence, the possibility of cortical dysfunction in autism has stimulated investigation. Specific pathophysiology of the temporal lobes and dysfunction of mesolimbic cortex and associated neostriatal structures have been postulated. Some quantitative EEG studies, both during wakefulness, without and with tasks, and sleep, suggest abnormal patterns of cerebral lateralization. These findings are not consistently supported by dichotic listening studies or CAT scans. CAT scans have shown abnormal structural configurations in only about one-quarter of autistic subjects, suggesting a subgroup in which the autism is associated with a structural brain abnormality. Impaired information processing is suggested by reports of small P300 waves in the evoked EEG response to target stimuli. Other event-related potential studies have failed to demonstrate cortical abnormalities.

Neurophysiological studies of subcortical events are relevant to the autistic disturbances of sensory modulation and motility. These investigations have included autonomic, vestibular, and brain stem auditory evoked response studies. The autonomic response studies have focused on the regulation of cardiovascular and respiratory responses mediated by the vagus nerve and originating at its source within the brain stem. It has been proposed that the increased heart rate variability of autistic children may reflect reticular formation responses to insignificant stimuli. Also, increased heart rate variability is greatest during autistic stereotypic behavior, linking dysmodulation of autonomic responsivity to the motility disturbances. Failure to habituate respiratory responses and enhancement of vascular responses to visual stimuli, indicating incapacity to reduce stimulus novelty and therefore sensory overload, link the abnormal autonomic responses to the disturbances of sensory modulation. Peripheral blood flow and heart rate are also elevated. The increased reactivity of autonomic responses may reflect the inability to gate or filter trivial sensory stimuli, thereby compromising appropriate selective attention. The vestibular response studies have demonstrated abnormal visual-vestibular interactions, prolonged time constants, and reduced secondary nystagmus. Abnormalities of vestibular adaptation and the influence of excessive reverberation in multisynaptic brain stem pathways on vestibular function have been proposed to account for these findings. Both the vestibular and autonomic responses distinguish autistic from control populations. Brain stem auditory evoked response studies, on the other hand, have not consistently supported the brain stem hypothesis; prolonged brain stem transmission times are found only in a minority of autistic children. The vestibular and autonomic responses probably involve widespread interconnecting neuronal fields within the brain stem. The mechanism underlying the autistic behavioral syndrome is likely to involve a system dysfunction rather than a pathologic change in a specific group of neurons. Brain stem auditory evoked responses are responses of a subset of neurons within the brain stem. It has been proposed that those autistics who do have prolonged brain stem transmission times may have brain stem neuropathology which replicates or activates the system dysfunction reflected by the vestibular and autonomic abnormalities.

In an attempt to integrate the clinical and experimental evidence for both cortical and subcortical neurophysiological dysfunction, it has been proposed that the disturbances of sensory modulation and motility reflect the pathophysiological mechanism and that the other abnormal behaviors can be understood as consequences of distorted sensory input. This suggests a neurophysiological dysfunction involving a cascading series of interacting neuronal loops in the brain stem and diencephalon which subserve modulation of sensory input. Some of the same systems modulate motor output in response to sensory input. The rostral projections from these structures include those to the hippocampus, limbic cortex, neostriatum, and parietal cortex. Parietal structures influence the direction of attention to stimuli of emotional significance, integrating sensory input from association cortex and thalamic centers with limbic and reticular input. Output is to regions involved with motor responses to emotionally significant stimuli. Autistic children suffer from distortions of directed attention and fail to sustain motor responses such as eye contact to socially relevant stimuli. Their emotional, cognitive, and language deficits suggest failure of emotional surveillance, a function attributed to dorsal parietofrontal structures. Brain stem and diencephalic centers project rostrally to telencephalic structures and these, in turn, modify brain stem and diencephalic function. In this model, hypotheses of rostrally and caudally directed sequences of pathoneurophysiological dysfunction in autism merge, so that autism can be explained in terms of dysfunction of brain stem and related diencephalic behavioral systems and their elaboration and refinement by selected higher neural structures.

Further reading

Ornitz EM (1983): The functional neuroanatomy of infantile autism. *Intl J Neurosci* 19:85–124

Ornitz EM (1985): Neurophysiology of infantile autism. *J Am Acad Child Psychiat* 24:251–262

Rutter M, Schopler E, eds. (1978): *Autism: A Reappraisal of Concepts and Treatment.* New York: Plenum Press

Behavior Therapy, Applied Behavior Analysis, and Behavior Modification

Leo J. Reyna

Behavior Therapy, Applied Behavior Analysis, and Behavior Modification refer to a relatively recent, interrelated group of procedures for clinical assessment, intervention, and prevention. They are employed to relieve behavioral, emotional, and cognitive distress and disabilities, and to enhance the individual's functional repertoires. The first journals, *Behavior Research and Therapy, Journal of Applied Behavior Analysis, Journal of Behavior Therapy and Experimental Psychiatry,* and *Behavior Therapy,* have been followed by many others.

This group of clinical procedures emerged from, and continues to reflect, laboratory studies on the psychology of learning begun 100 years ago. Behavioral clinicians have developed analyses and techniques designed to clarify (1) how functional and dysfunctional behavioral repertoires are acquired, (2) how these repertoires are maintained, and (3) how they can be weakened or attenuated. Their laboratory legacy led behavioral clinicians to give primacy to observable, measurable, replicable, and verifiable environmental events and behaviors, and this feature has led to the largest body of data-based, controls-designed research published in the psychotherapeutic literature.

Much of contemporary research and applications are concerned with demonstrations that learning plays a major role in dysfunctional, as well as functional, repertoires, and that both classes of repertoires can be modified by learning and unlearning interventions. Further, increasingly, behavioral interventions have been found useful in the remediation and rehabilitation of dysfunctional behavioral and emotional repertoires that are sequelae or by-products of a number of conditions affecting the biological integrity of the organism (e.g., genetic, constitutional, nutritional, injury, disease, and various medical disorders affecting the biochemistry and structural features of the nervous system).

Although today the terms Behavior Therapy, Applied Behavior Analysis, and Behavior Modification are often used interchangeably, historically, they are derivations from two major experimental areas in the psychology of learning. Behavior Therapy stemmed from the investigations of Pavlov and the formulations of Hull and Guthrie, and the area is variously referred to as Pavlovian or classical or respondent conditioning. Behavior Modification and Applied Behavior analysis stemmed from the work of Thorndike and Skinner, and concerned instrumental learning or operant conditioning. Respondents are largely unconditioned and conditioned reflexes, so-called involuntary reactions, and are of special interest in the analysis of the components of emotional and motivational behaviors. Operants are behaviors (more often chains of behavior) that have a more direct effect on the external environment, are the so-called voluntary behaviors seen in speech (covert and overt) and in the broad range of motor skills involving coordination, locomotion, and manipulation. While in respondent conditioning, emphasis was on the analysis of how new, neutral, antecedent stimuli came to control reflexes, in operant conditioning, analysis focused on how new behaviors take form as a result of the different classes of environmental effects that followed emitted behaviors.

Today, just as in the area of basic research on learning in which a total account of behavior increasingly includes analysis of respondents and operants, so too clinical behavioral interventions increasingly address the respondent and operant components present in dysfunctional repertoires and utilize intervention procedures directed at changing both classes of behaviors. Further, while Behavior Therapy was first (Guthrie, Salter, Wolpe) employed almost exclusively for office and outpatient clients, and behavior modification for resident or institutionalized patients (Lindsley, Allyon, Azrin, Ferster, Krasner), the full spectrum of intervention techniques are increasingly employed across all populations and settings.

Behavioral assessment

Many common features characterize the behavioral clinician's approach to identifying patient/client repertoires. The initial behavioral assessment interview seeks to secure certain broad details of the patient/client's life history: family and personal interactions, living arrangements, education, school and work histories, medical problems, etc. Following this, information is gathered on seven additional areas:

1. The presenting problem. Specific examples from everyday living (rather than such generalities as "I feel depressed most of the time") are solicited from the patient/client. The delineation (and treatment) of specific problem behaviors, emotional reactions, and cognitive behaviors is preferred to determining which diagnostic category the patient/client fits. The use of diagnostic labels is sometimes iatrogenic and often implies the presence of features not discernible and omits other features that are discernible in the patient/client's repertoires.
2. Onset and development of problem. When in time, and under what circumstances.
3. Problem fluctuations. Fluctuations, since time of onset, in the severity or absence of the problem(s), and identification of accompanying life circumstances.
4. Interference by problem. How problem interferes with a range of daily required and desired activities (e.g., work, school, relationships, sexual functioning, health). To further assist the patient/client in specifying information needed in these first four areas, the behavioral clinician often offers the client a variety of published inventories and checklists (not tests).
5. Functional repertoires. The delineation of the patient/client's positive functional repertoires (a) to use in conjunction with intervention procedures and (b) to avoid or coun-

teract the therapist and client's viewing the client as being solely dysfunctional.

6. Client's goals in therapy. The delineation of the goals or objectives of therapy as expressed by clients (in preference to goals set by therapists).
7. Client's definition of progress. The securing of an explicit statement of progress indicators from the client, in which clients specifically identify what observable, measurable changes in their own repertoires must take place to fulfill their own criteria of progress.

Areas (5–7) are designed to minimize the development of inappropriate dependency behaviors on the therapist and therapy.

The role of measurement (frequency, intensity, latency, and duration) of dysfunctional behaviors is emphasized at every stage of the assessment/intervention process. The behavioral clinician often trains and enlists the help of the client as data-gatherer. From the onset, the patient/client, significant others, and affiliated care-givers are likely to participate in keeping a daily log of the problem behaviors. This daily log is designed to identify the settings or circumstances that occasion problem behaviors and feelings, the client's reactions to these settings, and to record the consequences or the effects on self and others of these behaviors. In many instances, special forms for keeping such records have been developed to assist client/observers to keep track of particular problem behaviors (e.g., J. Cautela's organic dysfunction survey schedules). For resident populations, two or more observers may be enlisted to ensure reliability of the observations of problem behaviors.

Methods of behavioral intervention

Initial (and ongoing) assessment and evaluation procedures are designed to provide a detailed analysis of the patient/client's range of behavioral, emotional, and cognitive repertoires. Some repertoires are appropriate and useful (functional) and some are inappropriate and aversive (dysfunctional) for relating to people (including themselves), places, situations, interactions, things, tasks, knowledge, conceptions, ideas.

This discussion is primarily directed to the use of behavioral interventions with patient/client populations seen in office settings and for whom physical and medical disorders are not major factors contributing to their dysfunctional behavioral repertoires. Nevertheless, the intervention procedures are often useful as adjunctive procedures to somatic and medication interventions accompanying medical and physical rehabilitation procedures.

At least 30 different behavioral procedures for altering repertoires have been culled from the published literature (E. Reese) and since often more than one procedure is employed, the possible combinations of distinct components yields an even greater number of behavioral treatment programs.

For individuals whose dysfunctional behavioral repertoires consist predominantly of one or more high-frequency aversive emotional respondents elicited by particular environmental events (people, situations, thoughts, performing some task, medical disabilities, etc.), procedures for attenuating conditioned emotional reflexes include a variety of counterconditioning paradigms found, for example, in systematic desensitization, controlled relaxation, flooding, assertiveness training, modeling, extinction, and several others (used alone or in combination with each other and sometimes in combination with procedures described below).

Some behavioral clinicians (applied behavior analysts), however, may prefer to focus on the concurrent operants of avoidance and escape present with emotional respondents, and employ instead, a variety of reinforcement procedures for strengthening alternative or incompatible behaviors (i.e., approach repertoires to previously feared or discomforting events). Other clinicians, cognitive-behavioral, prefer to focus on, and alter high-frequency irrational thinking and dysfunctional self-statements (covert speech behaviors).

For individuals whose repertoires consist of one or more low strength (or absent) functional repertoires, intervention procedures may include one or more of the following procedures (partial listing): various positive reinforcement or contingent strengthening of functional behaviors, self-control and self-management procedures, shaping through reinforcement of successive approximation of the functional behaviors.

Since any particular dysfunctional repertoire includes a range of both high-strength dysfunctional behaviors, emotional reactions, and cognitions, and low-strength functional repertoires, several different intervention procedures are likely to be utilized for the same individual. Further, depending on the particular individual's history (personality), some patient/clients will be more responsive to a combination of particular methods for altering their behaviors.

Research

Earlier, the major emphasis in behavioral clinical research concentrated on studies of the efficacy of specific single procedures for specific dysfunctional behavior repertoires, but increasingly research is shifting to analysis of what combinations of procedures are most effective, and which components of a program are making which contribution to the total therapeutic outcome. Thus, the current intensive range of research in this area seeks to determine what specific procedures, alone or in combination, are most effective (immediate and long-term), in the shortest possible time, and with the least disruption (side effects) of daily activities, for different kinds of problems, presented by individuals with their own unique predysfunctional histories, personalities, life-styles, and settings. Reports of such research can be found in the Journals mentioned at the beginning of this article, as well as in three major research review series: *Annual Review of Behavior Therapy; Progress in Behavior Modification;* and *Advances in Cognitive-Behavioral Research and Therapy.*

Further reading

Hammonds B, ed. (1984): Psychology and learning. In: *The Master Lecture Series,* Vol 4. Washington DC: American Psychological Association

Skinner BF (1953): *Science and Human Behavior.* New York: Macmillan

Wilson G, Franks C, eds (1982): *Contemporary Behavior Therapy.* New York: Guilford Press

Wolpe J, Salter A, Reyna L, eds (1964): *The Conditioning Therapies: The Challenge in Psychotherapy.* New York: Holt, Rhinehart and Winston

Behavioral Medicine

Neal E. Miller

Behavioral medicine is an interdisciplinary field integrating basic research and applications from behavioral and biomedical sciences that are relevant to problems of physical medicine. The behavioral sciences involved are psychology, sociology, anthropology, and economics. In addition to the basic biomedical sciences, epidemiology and clinical observations are involved.

Although the behavioral sciences have had some relationship to medicine for a long time, primarily in the area of mental health and to some extent in the area of pediatrics, the explosive expansion of relationships with physical medicine is a phenomenon that took off in the late 1970s and is continuing to progress rapidly. One factor in the advance of behavioral medicine is an increased understanding of the various mechanisms by which the brain, via its neural and humoral systems, controls the health of the body.

Behavioral medicine is concerned with psychosocial factors across a wide range of medical activities: etiology, prevention, compliance with medical prescriptions, improving certain diagnoses and relieving the stressfulness of ominous ones, methods of reducing the stressfulness of aversive medical procedures and their aftereffects, direct or adjunctive therapy for certain physical conditions, rehabilitation, and problems of old age. Because of the great variety of activities in each of these areas, only a few illustrative examples are presented.

Smoking is a form of voluntary behavior clearly harmful to health. Converging lines of biomedical evidence show that cigarettes are causal factors in cardiovascular disease, cancer of the mouth, lungs, and esophagus, emphysema, bronchitis, and chronic obstructive lung disease. Behavioral studies show that nicotine is a strong reinforcer for smoking. Withdrawal symptoms and the difficulties that many people encounter in quitting indicate that smoking can be addictive. Some modest success in getting smokers to stop has been achieved by programs ranging from the use of the mass media to behavior-modification techniques, and even modest successes can produce considerable savings in the total burden of disease to be expected in the long run.

Since the problem of getting smokers to quit permanently is extremely difficult, efforts are being concentrated on prevention. After investigating the factors that cause children to start smoking, programs have been developed to teach young people strategies for counteracting pressures from peers, adults, and the media. Students are shown films of adolescents using various techniques to resist different inducements to smoke; then they participate in role-playing situations in which they resist such pressures. Because immediate rewards and punishments are so much more effective than delayed ones, rather than emphasizing remote dangers such as lung cancer, the programs emphasize the immediate consequences of smoking, such as increased heart rate, messiness, and odors. Follow-up studies two years later indicate that, among students exposed to such programs, the incidence of smoking is approximately half that of controls.

Alcohol abuse contributes to cirrhosis of the liver, pancreatitis, several types of cancer, and injuries from accidents and violence. Studies have shown that cultural and other psychosocial factors contribute to alcoholism, but we need to know far more about such causes. While a number of different social and behavioral programs are successful in treating some people, the average rate of long-term success is discouragingly low. This suggests emphasis on prevention. The amount of research devoted to alcoholism has been extremely small compared with the great social, economic, and medical importance of the problem.

Overeating and underexercise contribute to obesity, which increases the risk of hypertension, diabetes, and heart disease, and complicates surgery. Reduction in obesity can be a significant factor in treating certain cases of hypertension and diabetes. That some sort of influence in the social environment contributes to obesity is demonstrated by the fact that its incidence is much lower in people born into the upper social classes than those born into the lower ones. A wide variety of behavioral treatments can cause some obese people to return permanently to normal weight, but for most patients, relapse rates are discouragingly high. Many advances have been made in understanding various aspects of the complex mechanisms regulating food intake, but there is still no definitive way to reduce obesity rates.

If everyone complied with their physicians' recommendations, it would be easy to get rid of the unhealthy behaviors that have just been described. But this does not occur with recommendations about life-style or other problems. In fact, studies have shown that approximately one-third to one-half of patients do not take the drugs that are prescribed. Studies have shown that having the reason for the prescribed procedure thoroughly understood helps but often is not enough. Other studies have shown that compliance with recommendations of health care therapists is increased if patients have a relationship with them that is warm and affectionate and that bolsters the patients' self-confidence.

Clinical experience has been confirmed by behavioral experiments on animals proving that deprivation of stimulation by visual patterns during an early critical phase of development can produce a permanent deficit. Thus, the early correction of conditions such as severe astigmatism that interfere with clear patterned vision is important in order to avoid permanent deficits that cannot be corrected later by appropriate lenses. Behavioral research showing that infants have a strong tendency to look at a pattern rather than an unstructured part of the visual field has enabled the early diagnosis of defects in pattern vision and testing for the type of lenses that corrects the deficit.

Studies have shown that people exhibiting a combination

of hard-driving, competitive, impatient, and hostile behavior, which has been called type A, are approximately twice as likely to have heart attacks as those who do not show this pattern, defined as type B, even after other risk factors, such as smoking, obesity, and hypertension, are parceled out. One such study followed 3,500 people prospectively for 8.5 years. Another study, on a large sample studied as medical students showed that those scoring high on a scale measuring hostility and suspiciousness had substantially higher rates of mortality than those with low scores. Such studies are stimulating research into the personality variables involved in adverse medical consequences and the psychophysiological mechanisms that may be involved.

Clinical observations, epidemiological studies, and life-change scales strongly suggest that a variety of situations that can loosely be described as stressful increase the probability of a variety of adverse medical consequences. Although ingenious controls have been used, it is difficult to eliminate confounding factors, such as sanitation, personal hygiene, diet, and exercise. However, a number of rigorously controlled experiments, which necessarily could be conducted only on animals, have strongly supported the foregoing suggestive results by proving adverse effects of certain stressors on various aspects of the cardiovascular, gastrointestinal, and immune systems.

Studies have shown the interaction between psychological and organic factors. For example, when both groups are tested on a normal diet, the psychological stress of conflict induced by an electrical shock on the water bottle will produce hypertension in a population of rats selected to become hypertensive on a high salt diet, while it will not produce hypertension in rats selected to be resistant to such a diet. With dogs, when a stress of avoidance learning that would not by itself produce hypertension is combined with a moderate salt load that also would be ineffective by itself, the combination of the two produces hypertension. Similarly, when pigs or dogs are given experimental heart attacks by having a coronary artery tied off, a procedure that would not by itself lead to a fatal cardiac arrhythmia, sudden death can be induced by stimulating the sympathetic nervous system or by behavioral stressors. This fits in with the observation that when a group of doctors are making ward rounds through an Intensive Care Unit for patients recovering from heart attacks, the patients are five times as likely as at any other time of day to have arrhythmias producing fibrillation that would be fatal if equipment were not available to resuscitate them.

A specially significant analytical series of studies on rats has shown how purely psychological variables can affect the pathological consequences of identical levels of a physical stressor. For example, when the physical strength of electrical shock was controlled by having fixed electrodes on the tails of pairs of rats wired in series, a signal for one member of each pair that enabled it to learn the discrimination of when it was dangerous and when it was safe caused that group to develop one-fifth as many stomach lesions as the one without that signal. In another experiment, being able to control the shock by a simple coping response caused rats in that group to have fewer stomach lesions, lower levels of plasma corticosterone, and less depletion of brain norepinephrine than the rats that received identical shocks but were helplessly dependent on their partners.

The foregoing rigorously controlled experimental results are in line with the clinical observations that an individual's perception of a threat is more important than the objective danger and that the ability to have some form of control or to perform coping responses can be more important than the overall difficulty of the situation.

Yet other research shows how behavioral factors such as stress can have effects on the medically highly significant immune system. Finally, research using powerful new neurophysiological, biochemical, and pharmacologic techniques is providing increasing information about the brain mechanisms via which psychosocial factors can exert powerful effects on the health of the body.

On the other side of the picture, clinical and epidemiological evidence strongly suggests that positive emotional factors, such as social support, can contribute to freedom from disease and to longevity. But the effects of positive emotional factors have not yet been investigated by rigorous studies that control for possibly confounding factors. For example, people with many friends are probably more likely to be told that they need to see a doctor.

Studies of the effects of stress have provided the scientific rationale and impetus for the use of behavioral techniques to reduce stress as well as the use of other techniques such as inoculation against stress and training in various coping responses such as assertiveness training.

The techniques of behavior modification and behavior therapy, originally developed primarily to deal with the behavior of retarded children and neurotic patients, are increasingly being applied to a wide variety of problems of physical medicine, ranging from the prevention or retraining of unhealthy habits, such as smoking, through securing better compliance with medical prescriptions, to helping to cure conditions that have been difficult to treat by purely physical or pharmacologic approaches. An idea central to one important area of such application is that when sickness behavior is reinforced more strongly than healthy behavior, it will persist after the organic cause has disappeared. Sympathy and attention from family members and care-facility staff, release from onerous responsibilities and duties, the benefits of pain-killing or sleep-inducing medications, and payment for disabilities are powerful reinforcers for such behaviors. Some of the sickness behaviors reinforced in this way are signs of pain, such as wincing, limiting physical activity, asking for pain-killing drugs, symptoms of extreme dependence, fatigue, weakness, headaches, dizziness, and various other conditions that are either without an organic cause or that are disproportionate so that full use is not made of the patient's potential capabilities. In such cases, behavior therapy uses a variety of ingenious techniques to withhold reinforcements for sickness behavior and provide them for healthy behavior. This is done first in the health care situation and then extended to the patient's normal environment. In many cases, it may be necessary specifically to train the patient in healthy behavior as the way of achieving goals previously reached by sickness behavior. One of the best evaluated uses of behavior therapy has been in treating pain that has no discoverable organic basis. Phobias for procedures such as dental work, dialysis, injections, or having blood drawn have been eliminated by training patients to relax and then introducing them gradually to progressively more fear-inducing steps of the procedure along with demonstrations by models who are free of fear.

The autonomic nervous system, which controls the visceral organs and glands, had long been thought to be fundamentally inferior to the somatic one controlling the skeletal muscles. The strong traditional belief was that the autonomic nervous system could be modified only by classical conditioning, which was believed to be a primitive type of learning, and not by the more flexible type of trial-and-error learning (also called operant conditioning or instrumental learning) that is influenced by rewards and punishments and believed to be primarily responsible for voluntary behavior. Effects of the higher type of learning were thought to be indirect and hence trivial, as when blood pressure is increased as a by-product of muscular tension.

Relatively recent experiments on both animals and people, however, have shown that instrumental learning shaped by rewards and punishments can significantly modify a variety of visceral responses controlled by the autonomic nervous system, such as salivation, heart rate and rhythm, vasoconstriction and vasodilation, blood pressure, and the galvanic skin response. It is difficult to prove conclusively that the learned control is direct and not mediated by some skeletal response. But even in cases where the visceral response is clearly mediated by a skeletal one, clinical evidence shows that the effects are not trivial. For example, some patients for whom the symptom of tachycardia achieves the goal of escaping from some unwelcome type of responsibility produce their rapid heart rate by hyperventilation. Certain others, but unfortunately not all, who for organic reasons suffer sudden attacks or paroxysmal tachycardia can learn to abort such attacks by taking a sudden deep breath. Certain other cases with the arrhythmia of bigeminy (abnormal beats alternating with normal ones) can be taught to terminate this troublesome symptom by exercising to increase their heart rate. Persistent muscular tension can contribute to hypertension, and physical relaxation produces some relief.

The best evidence for direct control over an autonomically mediated response comes from patients paralyzed from the neck down by high spinal lesions. Such patients, who have suffered from orthostatic hypotension that confined them to a horizontal posture for two or more years and who resisted the conventional therapy of progressive elevation to more vertical postures on a tilt table, have been taught to achieve voluntary control over blood pressure by biofeedback training in which they are given moment-to-moment information about changes in blood pressure and rewarded for increasing it. This training has corrected their homeostatic defect of orthostatic hypotension and enabled them to achieve the much more normal activities that involve sitting upright. The large changes that such patients can learn to produce voluntarily without any observable changes in electromyograph recording from their nonparalyzed muscles or in their breathing are more than twice as great as those they can produce by obviously great effort at contracting as strongly as possible all their functional muscles and trying as hard as they can to contract all their paralyzed ones. They are also much greater than those that can be produced by hyper- or hypoventilation or changes in intrathoracic pressure produced by a Valsalva maneuver. Although it is difficult to prove a negative, such tests seem to have ruled out all types of skeletal mediation.

The apparently direct control over visceral responses mediated by the autonomic nervous system opens up interesting new theoretical possibilities for the role of learning in normal homeostasis and in the etiology and therapy of certain visceral disorders. One use of biofeedback, in combination with autogenic training and relaxing imagery, has been teaching experimental subjects to control motion sickness induced by exposure to a rotating chair. Following up on the original research by Dr. Cowings at the NASA Ames Research Center, a study by the U.S. Air Force showed that of 50 members of air crew who were about to be grounded because all other methods had failed to cure their incapacitating air sickness in modern high-performance planes, such training restored 40 to flying duty. *Biofeedback,* in combination with other behavioral techniques, is especially useful in improving the rehabilitation of patients with neuromuscular disorders and with urinary and fecal incontinence. Because of the contribution of the latter two disorders to the need for care of the elderly in nursing homes, Faye Abdullah, the U.S. Deputy Surgeon General, has estimated that the widespread application of behavioral techniques developed for these two disorders could save more than $10 billion a year.

Further reading

Levy SM, ed (1982): *Biological Mediation of Behavior and Disease: Neoplasia.* New York: Elsevier Biomedical

Lindemann JE (1981): *Psychological and Behavioral Aspects of Physical Disability.* New York: Plenum Press

Matarazzo JD, Miller NE, Weiss SM, Herd JA, Weiss SM, eds (1984): *Behavioral Health: A Handbook of Health Enhancement and Disease Prevention.* New York: Wiley-Interscience

Melamed BG, Siegel LJ (1980): *Behavioral Medicine: Practical Applications in Health Care.* New York: Springer

Miller NE (1983): Behavioral medicine: Symbiosis between laboratory and clinic. *Ann Rev Psychol* 34:1–31

Miller NE, Brucker BS (1979): Learned large increases in blood pressure apparently independent of skeletal responses in patients paralyzed by spinal lesions. In: *Biofeedback and Self-Regulation,* Birbaumer N, Kimmel HD, eds. Hillsdale, NJ: Erlbaum Associates, pp 287–304

Pomerleau O, Brady JP, eds (1979): *Behavioral Medicine: Theory and Practice.* Baltimore: Williams & Wilkins

Cocaine

Roger D. Weiss

Cocaine (benzoylmethylecgonine) is an alkaloid derived from the leaves of *Erythroxylon coca,* a shrub found in the eastern highlands of the Andes Mountains. Coca leaves, which contain cocaine in concentrations ranging from 0.6% to 1.8%, have been chewed by the inhabitants of this region for medicinal, religious, and work-related purposes for at least 1,500 years. The local anesthetic properties of cocaine were initially described by Sigmund Freud and Karl Koller in 1884; the drug is still widely used for topical anesthesia of the upper respiratory tract. However, the most common reason for cocaine use today is recreational; the drug is a central nervous system stimulant and a powerful euphoriant. It is because of this property that cocaine abuse has become widespread.

History

Coca leaves have been used since at least A.D. 500, when several bags of the leaves were found at a Peruvian grave site along with other items considered necessities for the afterlife. After the Spaniards conquered the Incas in the 16th century, they set aside their religious objections to the use of coca after seeing how chewing the leaves enabled the Indians to work long hours in gold and silver mines with little need for food or sleep. Indeed, coca leaves were eventually used by the Spaniards as a form of payment to the Incas.

Cocaine was first extracted from the coca leaf in 1855; this discovery was followed by a flourish of experimentation with the compound, peaking in the latter part of the 19th century. Freud noted the drug's ability to relieve pain from peripheral lesions and thus led the way to the discovery of cocaine as a topical and local anesthetic. He also claimed that cocaine might prove useful as a stimulant, as an aphrodisiac, and in the treatment of depression, gastrointestinal disturbances, wasting diseases, alcoholism, morphine addiction, and asthma. His work on cocaine was highly controversial, however, and he was accused of irresponsibility by much of the scientific community. Despite this, the general enthusiasm for cocaine led to its incorporation into a number of tonics and patent medicines including Coca Cola. The cocaine was removed from Coca Cola in 1903, but even now it is flavored by decocainized coca leaves.

Cocaine's growing popularity was accompanied by increasing concern over the ability of the drug to cause dependence. As more people abused the drug and deaths were reported from cocaine toxicity, restrictions were placed on cocaine use. In 1914, the Federal Harrison Narcotics Act, which incorrectly classified cocaine as a narcotic rather than as a stimulant, prohibited the use of cocaine in patent medicines and effectively made recreational use of cocaine illegal. For years the drug went underground and was used by a very small segment of the population. The growing availability of amphetamine, which was pharmacologically quite similar to cocaine, much less expensive, and legally obtainable through prescriptions, helped lead potential stimulant abusers toward that drug rather than cocaine. In the late 1960s and 1970s, greatly increased interest in and acceptance of recreational drug use by certain segments of the population contributed to the renaissance of interest in cocaine. Cocaine's recent surge in popularity has been abetted by its reputation as the prestigious "champagne of drugs."

The most common method of cocaine use consists of "snorting" crystalline cocaine hydrochloride intranasally. The drug is sold illicitly as a translucent white powder, usually mixed with adulterants such as other local anesthetics, sugars, and amphetamine. It is arranged into thin lines 2–5 cm long and 0.25 cm wide, and inhaled through a straw or a rolled-up dollar bill. An average line consists of 10 to 15 mg of cocaine. Intravenous use, though unusual in occasional users, is often seen in severe cocaine abusers. In this form of drug use, heroin is often combined with cocaine to form a "speedball"; simultaneous injection of the two drugs reduces the intensity of the cocaine-induced stimulation, which some users find uncomfortable. "Freebasing" refers to the smoking of the alkaline precursor (freebase) of cocaine hydrochloride. Cocaine freebase is prepared by dissolving cocaine hydrochloride in water, and then adding a base such as ammonia or baking soda. The alkaline precipitate which is thus formed can then be smoked, usually in a glass waterpipe.

A recently developed form of freebase, known as "crack," is cocaine freebase that has been manufactured from cocaine hydrochloride by the dealer rather than the user. The resultant product consists of small chips that resemble white pebbles; these rocks of ready-to-smoke freebase may be sold for as little as $10 each, and may contain cocaine in excess of 75% purity. The low price, ease of use, high purity, and extremely addictive nature of crack have caused great concern and fear of an epidemic of cocaine addiction.

Drug effects

Like other local anesthetics, cocaine is believed to inhibit nerve impulses by altering the nerve cell membrane. It has been posited that molecules of the anesthetic dissolve in the liquid matrix of the neuronal membrane and bind to receptor sites within sodium channels. This process interferes with the opening of the channels and thereby inhibits the transfer of sodium ions across the cell membrane. Depolarization of the axon cannot occur, and the nerve impulse is thus blocked. An additional property of cocaine which makes it particularly useful as a local anesthetic is its ability to cause vasoconstriction. Decreased blood flow in the anesthetized area allows for increased visibility during surgical procedures performed with cocaine anesthesia. Cocaine's vasoconstrictor property is particularly useful in otolaryngology, because of the high vascularity of the nose and throat.

Although it was once felt that the stimulant effects of cocaine occurred as a result of the drug's ability to increase norepinephrine turnover in the central nervous system, more recent evidence has suggested that cocaine causes euphoria primarily by prolonging the action of dopamine at nerve junctions where it serves as a neurotransmitter. This augmentation of dopaminergic activity, which takes place most prominently in the nucleus accumbens in animals, results from blockade of dopamine reuptake.

Following intranasal administration of cocaine, onset of action begins within 2 minutes. Intravenous use and freebase smoking produce almost immediate effects. Symptoms of intoxication last for 20 to 40 minutes after intranasal use, and half that long after intravenous injection or freebase smoking. In general, the subjective effects of cocaine are related not only to the plasma level achieved but the rate of rise of plasma level. The drug is metabolized in the liver and by cholinesterase enzymes in plasma. Cocaine is eliminated following first-order pharmacokinetics, with a half-life of approximately 1 hour. The metabolite benzoylecgonine is excreted in the urine and can be detected by enzyme immunoassay up to 72 hours after the last dose.

Clinically, the most prominent effect of cocaine on mood is one of euphoria. Stimulation, reduced fatigue, increased mental clarity, diminished appetite, and garrulousness are also frequently reported consequences of cocaine use. Physical effects include tachycardia, systolic and diastolic hypertension, increased body temperature, and dilated pupils. The euphoriant effects of cocaine may lead some vulnerable individuals to abuse the drug chronically. Animal studies have shown that cocaine is an extremely powerful reinforcer; rhesus monkeys given unlimited access to intravenous cocaine will repeatedly self-administer the drug to the point of severe toxicity and even death. Repetitive use in humans may lead to sleeplessness, loss of appetite, severe anxiety, paranoia, delusions of persecution, and auditory, visual, and tactile hallucinations. Chronic cocaine abusers often spend large sums of money in an attempt to obtain the drug and may resort to illegal activity in this pursuit. Some abusers have been known to spend up to $2,000 a day on cocaine. Chronic intranasal use may lead to inflammation or ulceration of the nasal mucosa, and chronic freebase smoking may cause pulmonary dysfunction. Intravenous users are exposed to the medical hazards of unsterile needles, including abscesses, hepatitis, infective endocarditis, and AIDS. Efforts to identify those at risk for chronic cocaine abuse have begun, though little is known at this time about specific characteristics which may predispose individuals to develop this disorder.

Further reading

Fischman MW, et al. (1976): Cardiovascular and subjective effects of intravenous cocaine administration in humans. *Arch Gen Psychiatry* 33:983–989

Siegel RK (1977): Cocaine: recreational use and intoxication. In: *Cocaine 1977*, Petersen RC, Stillman RC, eds. Washington DC: National Institute on Drug Abuse research monograph 13

Weiss RD, Mirin SM (1984): Drug, host and environmental factors in the development of chronic cocaine abuse. In: *Substance Abuse and Psychopathology*, Mirin SM, ed. Washington DC: American Psychiatric Press

Weiss RD, Mirin SM (1987): *Cocaine* Washington DC: American Psychiatric Press

Wise R (1984): Neural Mechanisms of the reinforcing action of cocaine. In: *Cocaine: Use and Abuse*, Grabowsky J, ed. Rockville Md: National Institute on Drug Abuse

Coma

Fred Plum and Harriet O. Kotsoris

Coma and clinical states allied to it represent severe or total global reductions of mental function due to structural or physiological impairments of the brain. *Coma* is characterized by a sustained loss of the capacity of arousal, preventing any expression of potential mental function. The eyes are closed, sleep-wake cycles disappear, and even vigorous stimulation produces no evidence of appropriate psychological response. *Stupor* describes a state of spontaneous unarousability in which strong external stimuli can transiently restore wakefulness. Stupor implies that evidence of at least a limited degree of appropriate mental activity accompanies the arousal, transient though it may be. *The vegetative state* is the condition wherein arousal (i.e., sleep-wake cycles) returns or remains but appropriate testing measures elicit no evidence of the person's cognitive awareness of self or environment. *The locked-in state* describes a condition in which persons retain or regain arousability and self-awareness but because of extensive bilateral paralysis (i.e., deafferentation) can no longer communicate except in severely limited ways. *Brain death* describes the irreversible loss of function of all neural structures rostral to the foramen magnum.

Mechanisms

Eyes-closed coma always arises as a result of an acute disorder that produces sudden depression or damage of specific and nonspecific diencephalic and upper brainstem arousal mechanisms. Conditions included in four major pathologic groupings can cause such severe, global acute reductions of consciousness. (1) In diffuse or extensive multifocal bilateral dysfunction of the cerebral cortex, the cortical gray matter is diffusely and acutely destroyed or depressed. Examples include laminar cortical depression or necrosis from severe hypoxia-ischemia, similar damage from sustained profound hypoglycemia, and metabolic depression from certain stages of hepatic encephalopathy. As a result, cortical-subcortical physiological feedback loops are impaired with the result that brainstem autonomic arousal mechanisms become profoundly inhibited, producing the equivalent of acute reticular shock below the level of the lesion. (2) Direct damage to paramedian upper brainstem and posterior-inferior diencephalic ascending arousal systems blocks normal cortical activation. Examples of conditions producing such damage include midbrain infarcts or the inflammatory lesions of certain selective forms of encephalitis. Anatomically, the affected structures lie predominantly in the paramedian gray matter extending roughly from the level of the nucleus parabrachialis of the pontine tegmentum forward as far as the ventral posterior hypothalamus and adjacent pretectal area. (3) Widespread disconnection between the cortex and subcortical activating mechanisms acts pathophysiologically to produce effects similar to both conditions 1 and 2 above. A typical example occurs with acute, severe damage to the cerebral hemispheric white matter (dysmyelination), a circumstance that occurs following periods of prolonged cerebral hypoperfusion-hypoxemia, e.g., due to carbon monoxide poisoning. (4) Diffuse disorders, usually metabolic in origin (metabolic encephalopathy), concurrently affect both the cortical and subcortical arousal mechanisms, although individually to a different degree.

As noted, coma reflects acute or subacute but always profound involvement of one or more of these four mechanisms. Acute abnormalities that are similarly distributed but less intense cause delirium, a condition marked by fluctuating levels of attention and arousal associated with impaired memory, confusion, distractibility, and, sometimes, delusions or hallucinations, especially visual ones. More insidiously developing and chronically lasting disorders of cortex and subcortex usually spare arousal mechanisms and result in a gradual decline in cognitive functions (dementia) without producing clouded consciousness.

Nearly all patients who survive coma sooner or later reawaken, with the interval between the acute damage and the return of wake-sleep cycles being longest following extensive

Table 1. Major Causes of Stupor and Coma

Supratentorial lesions (secondarily causing deep diencephalic upper brainstem dysfunction)
Cerebral hemorrhage
Large cerebral infarction
Subdural hematoma
Epidural hematoma
Brain tumor
Brain abscess (rare)
Subtentorial lesions (compressing or destroying the reticular formation)
Pontine or cerebellar hemorrhage
Brainstem infarction
Brainstem or cerebellar expanding tumor
Cerebellar abscess
Metabolic and diffuse lesions
Global cerebral anoxia or ischemia (e.g., cardiac arrest)
Hypoglycemia
Severe nutritional deficiency (e.g., advanced Wernicke's disease)
Endogenous organ failure or deficiency (e.g., lung, liver, kidney)
Exogenous poison (e.g., alcohol, sedatives, opiates)
Infections
Meningitis
Encephalitis
Ionic and electrolyte disorders (e.g., hyponatremia, water intoxication)
Status epilepticus
Concussion and postictal states
Psychogenic unresponsiveness

diencephalic-midbrain damage or when repeated severe medical problems complicate convalescence. The degree and quality of associated cognitive recovery once the effects of acute injury-disease pass depend upon how much residual damage involves cerebral cortical areas and the nonspecific subcortical nuclei that interact with limbic and association cortex. Metabolic suppression of the brain or limited degrees of structural damage such as occur with concussion can be followed relatively promptly by a complete or near complete return of brain function. More severe injury, especially to the cerebral cortical mantle or its specific and nonspecific relay nuclei leaves in its wake increasing degrees of intellectual impairment. Complete or near complete permanent decortication such as occurs with profound hypoxia, hypoglycemia, or sometimes with closed-head injury produces the horror of the vegetative state: a body that if externally cared for will maintain its internal homeostasis indefinitely but can no longer perceive, feel, think, or communicate.

Etiology

Coma implies the imminent threat of brain failure with widespread loss of cerebral function, upper brainstem function, or both. Table 1 lists the major diseases or categories that can produce such severe impairments. Supratentorial mass lesions generally cause relatively little loss of consciousness unless they expand and distort the brain sufficiently to compress the diencephalon caudally and produce transtentorial distortion or herniation. Upper brainstem–lower diencephalic lesions, by contrast, cause coma when lesions directly damage the central ascending activating systems. Metabolic disorders can affect both the supra- and subtentorial mechanisms that normally generate conscious behavior; accordingly, they generally produce commensurately widespread symptoms and signs of neurological dysfunction.

Except in overdose by sedative drugs, almost all examples of which recover with adequate treatment, brain dysfunction sufficient to produce coma implies a poor prognosis. Among large series of patients with nontraumatic coma lasting more than 12 hours or so, only about 15% completely recover their physical and intellectual functions. Likewise, among patients in sustained coma from head injury, almost half will die and as many as 25% of the survivors will be severely incapacitated. Patients with metabolic coma or young persons unconscious from head trauma generally do best, while those showing signs of severe primary or secondary brainstem damage fare worst no matter what the cause.

Patients in coma are best treated acutely in special care units where they can receive close attention to their often precarious autonomic functions and can be given specific therapy directed at their underlying neurological disease. Respiratory and cardiovascular support provide the mainstays with specific measures directed at the particular disease that threatens to destroy the brain.

Further reading

Jennett B, Teasdale G, Braakman R, et al (1978): Prognosis of patients with severe head injury. *Neurosurgery* 4:283–288

Levy DE, Bates D, Caronna JJ, et al (1981): Prognosis in nontraumatic coma. *Ann Int Med* 94:293–298

Plum F, Posner JB (1982): *The Diagnosis of Stupor and Coma*. 3rd rev ed. Philadelphia: FA Davis

Convulsive Therapy

Max Fink

Convulsive therapy is a psychiatric treatment for patients with major affective disorders. The treatments seek to change mood, affect, and interpersonal behavior by altering brain functions by a series of grand mal seizures. A successful course requires from 8 to 12 treatments, spaced 48 to 72 hours apart. Seizures may be induced by chemical or electrical means with equal efficacy. The principal benefits are the rapid relief of disordered mood, improvement in the vegetative functions, and reduction in suicidal drive. The principal risks are the development of an organic mental syndrome, usually cognitive impairment (memory loss), and fracture. While 8% of patients admitted to academic in-patient psychiatric services in the U.S. receive electroconvulsive therapy (ECT), the treatments are infrequently used in municipal, state, and federal facilities. About 100,000 patients are treated annually.

History

Convulsive therapy was developed in 1933 by the Hungarian neuropsychiatrist Ladislas Meduna, following reports that epilepsy and schizophrenia were rarely found in the same patients, suggesting a biological antagonism between the two disorders. These data were supported by his studies of the concentration of glia in postmortem brain samples, in which he found a diminution of glia in the brains of patients with dementia praecox and an overabundance in the brains of epileptics. The first patient was treated on January 24, 1934, using intramuscular camphor injections to induce the seizures. In 1938, the Italian psychiatrists Cerletti and Bini demonstrated the merits of alternating electrical currents through bitemporal electrodes as stimulating currents. Trials of brief pulse stimuli followed, and in the past decade both current forms have been used. In the mid-1950s, inductions using the inhalant flurothyl (Indoklon) were found to be equally effective, but more cumbersome. In 1961, the safety of seizures through unilateral electrode placements over the nondominant hemisphere was demonstrated.

Indications

ECT is recommended for patients with severe depressive disorders requiring hospitalization. The presence of vegetative symptoms (anorexia, weight loss, insomnia, loss of libido, impaired concentration, amenorrhea) is a good prognostic sign. Severe inanation, suicidal preoccupations and drive, and catatonic symptoms are indications for its use. While depressed patients with psychotic features (delusional depression) usually fail to respond to tricyclic antidepressants alone, they respond rapidly to ECT. In direct comparisons in random assignment studies, convulsive therapy is more effective for major depressive disorders than sham ECT, anesthesia alone, tricyclic antidepressants, or monoamine oxidase inhibitors.

Convulsive therapy is also recommended in patients with depressive or manic disorders who have not responded to other treatments. It is not recommended for patients with neuroses, anxiety, psychopathy, drug dependence, or character disorders. It remains effective in schizophrenia, but is infrequently used, since antipsychotic drugs are believed to be easier to use.

Procedures

Prior to ECT, the treatment procedures are explained in detail and consent obtained from patients and nearest kin. Special considerations for consent apply in patients who may be hospitalized involuntarily.

An intravenous line is established; methohexital and succinylcholine administered; an airway established; and the patient ventilated with 100% oxygen. Electrodes are usually applied over the nondominant hemisphere, and current dosages are defined as slightly above the minimum necessary to achieve a seizure. Seizure duration is monitored by the electroencephalogram (EEG), the motor activity in a limb isolated from succinylcholine by a blood pressure cuff inflated above the systolic pressure, or by heart rate. Seizure durations less than 25 seconds are considered inadequate and are usually repeated. Treatments are given either three or two times a week, the number of treatments determined by clinical signs of relief. Treatment courses range from 6 to 15 treatments, with an average of 7 to 9.

Treatments are given either with ac instruments, delivering between 50 and 150 J; or brief pulse instruments, delivering currents about one-third in intensity. Efficacy is related to current intensity and current path, so that some therapists prefer ac currents and bilateral electrode placements, especially in the more severely ill patients. But the use of these currents and pathways is associated with greater degrees of cognitive impairment.

Following a successful course, maintenance treatment is recommended, especially in patients with a history of recidivism. Maintenance is accomplished with tricyclic antidepressants, monoamine oxidase inhibitors, lithium, or the combination of a tricyclic antidepressant and an antipsychotic agent in patients with delusional depression. Maintenance ECT is also used.

Consequences

Fracture, organic mental syndrome, and cardiovascular complications leading to death are the principal risks. Fractures are rare under present conditions of anesthesia; as is death, which occurred in 4 of 100,000 treatments. Cognitive impairment, seen as persistent complaints of memory loss, is reported in about 1% of patients. The incidence is greater in patients treated with ac instruments and bilateral electrode placements. Memory of recent events around the period of hospitalization and the acute illness are most disrupted. Neither remote memory nor memory of events after the treatment course are affected.

The principal behavioral changes are a heightening of mood and affect, relief of depressive mood and suicidal ideation, and in patients with manic states, relief of the elevated mood. Delusional ideation gradually disappears. Motor excitement, withdrawal, and apathy are relieved, and vegetative functions improve.

Neuropsychological tests, including memory, motor functions, reaction time, and perception, are progressively impaired during the treatment course, but recover within four to six weeks after the last seizure. In patients who recover from their depressive illness, there is an improvement in performance and cognitive tests over pretreatment values.

During each seizure, the EEG exhibits high-voltage, high-frequency spike activity associated with slow wave bursts. In interseizure records, frequencies slow, amplitudes increase, and fast frequencies are reduced. Following treatment, slow-wave frequencies are reduced and synchronized alpha activity increases. During each seizure, there is transient hypertension, tachycardia, and, in many subjects, arrhythmias. Hematopoietic changes are minimal.

With each seizure, there is a transient increase in cerebral catecholamines, acetylcholine, proteins, and peptides, including prolactin, vasopressin, adrenocorticotropic hormone, and thyrotropin releasing hormone (TRH) with a return to baseline levels within two hours. Increases in cholinesterase and monoamine oxidase in the cerebrospinal fluid and a reduction of calcium in plasma, cerebrospinal fluid, and blood may persist. Turnover of catecholamines and receptor sensitivity increase, particularly dopamine synaptic activity. In patients with evidence of neuroendocrine dysregulation (as abnormal cortisol and failure to suppress with dexamethasone, abnormal growth hormone response to apomorphine, or thyrotropin-stimulating hormone response to TRH), there is a normalization of these responses with treatment.

Permeability of the blood brain barrier increases with each seizure, and this increase is usually persistent after a course of seizures. Neuropathologic examination of animals subjected to repeated seizures fail to show evidence of cellular dysfunction, either acute or chronic, unless seizures have been prolonged and accompanied by impairment of ventilation and reduction in oxygen saturation of the blood.

Mode of action

Present theories parallel those of the action of tricyclic substances, emphasizing increased turnover and levels of catecholamines, and increased adrenergic or dopamine receptor sensitivity. Alternate hypotheses emphasize diencephalic effects and focus on the probability that peptides with behavioral effects are released in increased amounts from hypothalamic centers during seizures. These hypotheses usually remark on the increased permeability of the blood brain barrier and the intracellular movement of calcium ions as associated phenomena in the therapeutic chain.

Further reading

Fink M, Kety S, McGaugh J, Williams T, eds (1974): *Psychobiology of Convulsive Therapy*. Washington DC: VH Winston & Sons

Fink M (1979): *Convulsive Therapy: Theory and Practice*. New York: Raven Press

Lerer B, Weiner RD, Belmaker RH, eds. (1984): *ECT: Basic Mechanisms*. London: John Libbey & Co

Malitz S, Sackeim HA, eds. (1986): *Electroconvulsive Therapy: Clinical and Basic Research Issues*. New York: New York Academy of Sciences

Weiner RD (1984): Does electroconvulsive therapy cause brain damage? *Behav Brain Sci* 7:1–53

Creutzfeldt-Jakob Disease

Clarence J. Gibbs, Jr., and *David M. Asher*

Creutzfeldt-Jakob disease (CJD), together with its clinical variants, is the more common of the two subacute human spongiform encephalopathies, the other being kuru. The spongiform encephalopathies are slow infections caused by filterable self-replicating agents, the nature of which is not yet fully elucidated. The human diseases resemble three conditions of animals: scrapie of sheep and goats, transmissible mink encephalopathy, and wasting disease of mule deer and elk.

Creutzfeldt-Jakob disease was first described in 1921 by Jakob, who attributed precedence to Creutzfeldt for the earlier report of a similar encephalopathy no longer accepted as being the same entity. The term spongiform encephalopathy was originally proposed by Nevin to describe an encephalopathy that he considered to be distinct from CJD, but the two terms came to be used almost interchangeably, and, following the suggestion of Gibbs and Gajdusek, spongiform encephalopathy became a generic term for the entire group of similar human and animal diseases.

Between 1921 and 1968 only 150 cases of CJD had been reported in the medical literature; more than 2,000 cases have now been recorded. The disease occurs throughout the world, mainly in patients of middle age, although rarely in younger people. Males and females are represented in approximately equal numbers. Estimated mortality rates vary from 0.25 to about 2 deaths per million population per year; there are several geographic foci of higher frequency, especially striking in Libyan Jews in Israel among whom the rate reached more than 30 per million per year. Most cases are sporadic, but CJD sometimes runs in families. In the United States about 10 percent of cases have a family history of presenile dementia.

The initial complaints in CJD are generally vague sensory disturbances especially involving vision, confusion, and inappropriate behavior, or, less commonly, cerebellar ataxia. Disease inevitably progresses over the following weeks or months to frank dementia and coma. Most patients have myoclonic jerking movements at some time during illness; some have convulsions. Spasticity or rigidity and decorticate primitive reflexes are present late in the disease. A variety of less constant findings, not necessarily symmetrical, occur, indicative of diffuse and multifocal degeneration throughout the brain in CJD. Clinical subtypes of CJD are sometimes given separate eponyms, e.g., Gerstmann-Straussler syndrome for familial CJD with cerebellar ataxia and amyloid plaques and Heidenhain variant for CJD with cortical blindness. The electroencephalogram in CJD often shows periodic "suppression-burst" complexes on a slow background. There may be moderate elevation of protein in the cerebrospinal fluid. As disease progresses cerebral atrophy with enlargement of the ventricles may be detected by computed tomography. Other laboratory studies are not useful in diagnosis. The mean survival of patients with CJD in several series is less than one year from onset, although about 10% of cases in our experience have lived for more than two years. Some cases of CJD, especially those of long duration, may be difficult to distinguish from the much more common Alzheimer's disease. Rare cases of CJD in young people may resemble subacute sclerosing panencephalitis, but that condition should be readily diagnosed by finding elevated levels of antibodies to measles virus in the cerebrospinal fluid.

The definitive diagnosis of Creutzfeldt-Jakob disease must still be made by histological examination of brain tissue obtained from biopsy or autopsy. The typical changes of spongiform encephalopathy are present, most pronounced in the cerebral gray matter: they include vacuolation of neurons leading to status spongiosus (Fig. 1), severe neuronal loss, and astroglial proliferation and hypertrophy. Amyloid plaques similar to those found in kuru have been reported in about 15% of cases of CJD. No consistent histopathological changes are recognized in organs outside the central nervous system (CNS), though the etiological agent has been detected in these tissues. The demonstration by negative-stain electron microscopy of unique helical fibrils (SAF, Fig. 2) in brains and spleens of patients and animals with the spongiform encephalopathies, and detection by gel electrophoresis and immunoblotting of abnormal low-molecular-weight proteins (PrP, Fig. 3) in those tissues may eventually prove useful in confirming histopathological diagnoses of CJD. Attempts are now in progress to demonstrate unique proteins of diagnostic significance in cerebrospinal fluid (CSF) of patients with CJD. Transmission of spongiform encephalopathy to susceptible animals by inoculation of suspensions of tissues from patients remains the ultimate

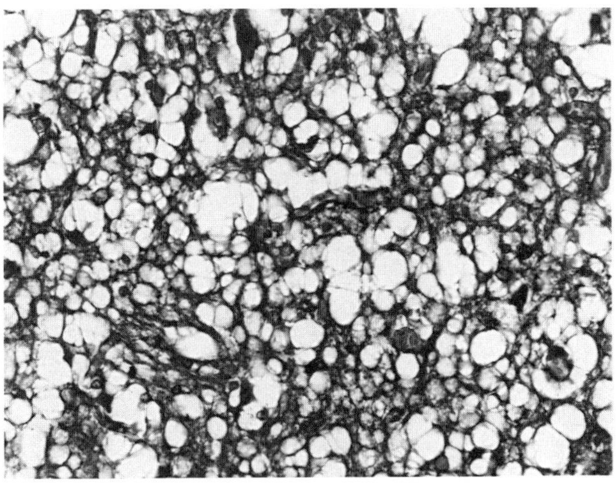

Figure 1. Severe status spongiosus in cerebral cortical gray matter of a patient with familial Creutzfeldt-Jakob disease. Hematoxylin and eosin, X330.

test for confirming the diagnosis of CJD. A variety of laboratory animals (hamsters, mice, guinea pigs) are susceptible to CJD, but in our experience, only chimpanzees and squirrel monkeys have been consistently susceptible with mean incubation periods of less than three years.

The physical structure of the etiological agent of CJD remains unknown, but its properties generally resemble those of the agent of scrapie. They are clearly subprotist in size and replicate in tissues of people and susceptible animals. Until recently agents of the spongiform encephalopathy group were usually called viruses, and still are by most authorities. However, they display a spectrum of resistances to inactivation by a variety of organic and inorganic chemicals and physical treatments, including both ultraviolet and ionizing irradiations, and heat, unknown among other viruses. That stimulated hypotheses that the spongiform encephalopathy agents might not be viruses at all, but unique pathogens containing no nucleic acids. Because the agents are sensitive to protein-denaturing treatments, and are therefore believed to contain proteins, Prusiner proposed that they be called *prions* for proteinaceous pathogens likely to be devoid of nucleic acids and subviral in size.

The prion hypothesis, however, has not been well substantiated or generally accepted. The aberrant behavior of the scrapie agent in chemical, heat, and irradiation inactivation-kinetic studies has been reinterpreted as consistent with that of a small virus, a resistant fraction of which is protected from inactivation by hydrophobic aggregation of its particles into masses as well as by association with host proteins. No confirmed filtration, sedimentation, or exclusion chromatographic studies have unequivocally demonstrated the transmissible scrapie particles to be subviral in size. No purified protein preparations free of nucleic acids have been found to retain scrapie infectivity. Should the transmissible agents of the spongiform encephalopathies eventually prove to be infectious proteins, whether encoded by derepressed host genes or by some novel mechanism through which heritable information is somehow inscribed in protein itself, the term prion would seem apt. If, however, the infectious particles of spongiform encephalopathy agents do contain nucleic acid genomes, then they might be better designated as unconventional viruses.

Three structures are now under investigation as representing the particles or important parts of the particles of the spongiform encephalopathy agents. The scrapie-associated fibrils (SAF, Fig. 2) were first described by Merz and colleagues in 1981: they appear as pairs or double pairs of helical structures that resemble but are distinguishable from cerebral amyloid fibrils of Alzheimer's disease. Prusiner and co-workers found abnormal low- molecular-weight sialoglycoproteins (PrP_{27-30}) in extracts of tissues from animals with scrapie (Fig. 3), and antigenically related proteins in patients with CJD. Those proteins are components of prion rods that appear to be virtually indistinguishable from Merz's SAF. Both SAF and Prp_{27-30} can be found in partially purified preparations of infectious scrapie agent. Fibrils containing antigenically related proteins have recently been described in thin sections of infected brains as well. It has been proposed that SAF/PrP represents the actual infectious particles of the spongiform encephalopathy agents or moieties or polymers of them. However, the demonstration that at least part of the PrP_{27-30} is encoded by a normal host gene, and that a larger antigenically related protein, presumably a precursor, is found in normal tissues, favors the possibility that it is a pathological host protein that has not been separated from the actual infectious particles by current techniques.

If the spongiform encephalopathy agents are small unconventional viruses, it is possible that they could be the small tubulovesicular structures, approximately 23 nm across, first observed in thin sections of scrapie-infected mouse brains by David-Ferreira, Gibbs, and co-workers in 1968, and subsequently identified in a variety of experimental and natural spongiform encephalopathies by Narang and others. These structures have yet to be identified in negatively stained suspensions of infectious material or otherwise characterized.

A second unresolved question concerns the natural mechanisms of transmission of CJD. It is important to identify these mechanisms; experience with kuru suggests that interrupting the major natural mechanism of transmission of a spongiform encephalopathy is possible by social means and served to end the epidemic of that disease. Epidemiological surveys have investigated several hypothetical mechanisms of CJD spread, including contamination of meat products with scrapie agent and iatrogenic transmission of the infectious agent from patients with the disease. The hypothesis of iatrogenic spread has been proved, since accidental transmission of CJD by contaminated corneal transplants and cortical electrodes has been convincingly documented. The occurrence of clusters of CJD in patients with a history of previous neurosurgery at

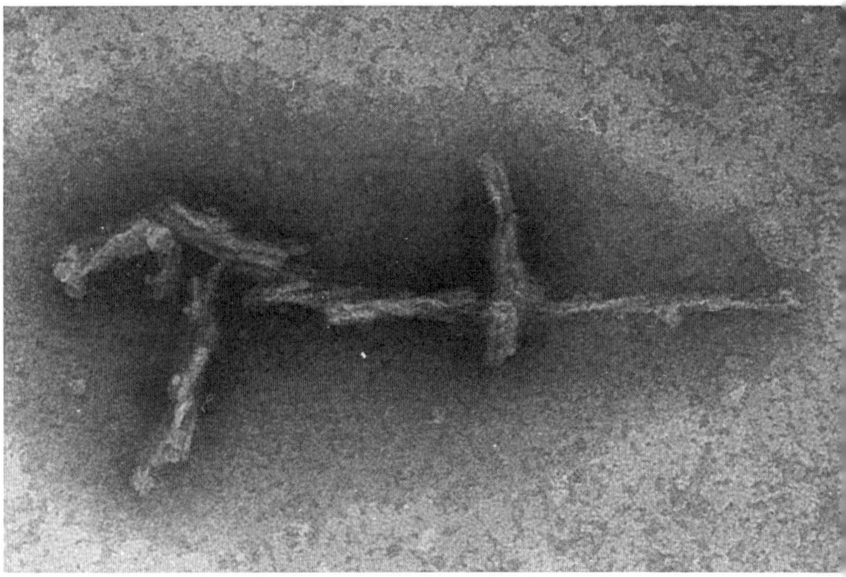

Figure 2. Scrapie-associated fibrils (SAF) in negatively stained extract of scrapie-infected hamster brain. Phosphotungstic acid, X144,000.

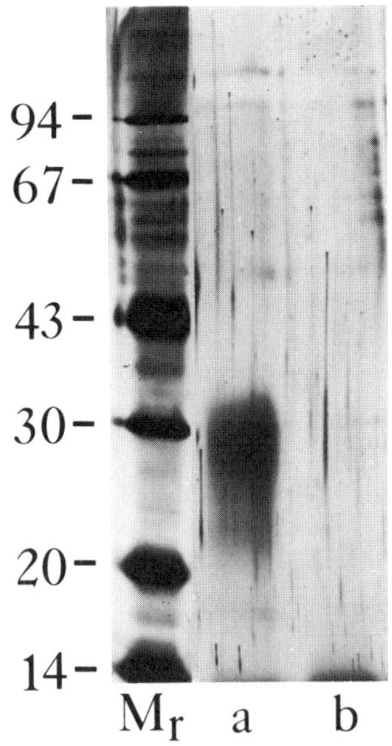

Figure 3. Low-molecular-weight protein band (PrP_{27-30}) in extract of scrapie-infected hamster brain. Sodium dodecyl sulfate–polyacrylamide gel electrophoresis. Silver stain. Mr, molecular-weight markers; a, scrapie-infected brain; b, control brain. Courtesy of Drs. M. Coker-Vann and P. Brown.

a common site and in young people who received injections of human pituitary growth hormone from a common source also implicated iatrogenic transmission. In the great majority of cases, however, the hypothetical iatrogenic events remain unknown; possible mechanisms awaiting investigation include ocular tonometry, surgery of head, neck, and teeth, and trauma surgery.

Familial CJD also remains incompletely understood. The pattern of occurrence in affected families roughly resembles an autosomal dominant mode of inheritance. However, familial CJD is otherwise clinically and histopathologically indistinguishable from sporadic CJD, and the infectious agent associated with both forms seems to be the same. It is not known if familiar CJD results from true genetic increased susceptibility to infection with an ubiquitous agent that is ordinarily of low transmissibility or from pseudogenetic person-to-person spread; experimental studies in animals have confirmed the existence of hereditary differences in apparent susceptibility to scrapie mediated by single-genes.

Spouses and other household contacts of patients seem to be at very low risk of acquiring CJD. Personnel caring for patients and handling their tissues are not known to have contracted the disease, although it has been diagnosed in physicians and other medical workers with possible occupational exposure. It is reassuring that, in the more than 60 years since CJD was first described, no pathologist has been diagnosed with CJD.

Brain tissues of patients with histologically confirmed CJD have been consistently infectious, containing from a thousand to more than a million lethal doses of agent per gram when assayed in susceptible animals. The infectious agent has been found less regularly and probably in lower amounts in several other tissues and in CSF; infectivity has not yet been demonstrated in blood or other body fluids, secretions, or excretions of CJD patients. However, because only a small number of such materials have been adequately tested, because blood from animals with experimental CJD has been found to contain the agent, and because the minimum amount of agent needed to infect human beings is not known, it is prudent to consider all tissues and excretions of patients as potentially contaminated. We advise using precautions for handling CJD patients and their materials similar to those for viral hepatitis.

Only three regimens are currently recommended for disinfecting the agent of CJD: heat, lye, and bleach. Heating, by incineration of disposable materials or steam autoclaving for at least one hour, should be employed whenever possible. Autoclaving at higher than usual temperatures (we recommend 132·C instead of the usual 121·C) is preferrable. One normal or higher concentration of sodium hydroxide has been effective in inactivating large amounts of both scrapie and CJD agents; we recommend exposing contaminated materials to hydroxide for at least one hour. Sodium hypochlorite (5.25%) (fresh, full-strength household chlorine bleach) has considerable inactivating potency in several experimental studies; at least one hour of exposure is recommended. Sterilization with ethylene oxide gas and a variety of commercial liquid disinfectants has been found ineffective and should not be attempted. The procedures employed for decontamination of materials potentially contaminated with the CJD agent must be adapted to the particular situation in each institution. It must be stressed that empirical evidence suggests that the major risk is of accidental transmission of CJD to other patients and not to the staff, and safety precautions must not be allowed to interfere with adequate diagnosis or humane care of demented patients.

There is still no effective therapy for CJD or the other spongiform encephalopathies. Several substances apparently interfered with the appearance of scrapie when administered before or shortly after animals were inoculated with low doses of the infectious agent, possibly by blocking replication of the agent in the reticuloendothelial system; none reversed established infection of the CNS. Occasionally remissions of CJD after various treatments have been claimed but never confirmed in subsequent attempts.

Further reading

Gajdusek DC (1977): Unconventional viruses and the origin and disappearance of kuru. *Science* 197:943–960

Gajdusek DC (1985): Unconventional viruses causing subacute spongiform encephalopathies. pp 1519–1557. In: "Virology" BN Fields, ed. New York: Raven Press

Merz PA, et al (1984): Infection-specific particle from the unconventional virus diseases. *Science* 225:437–440

Oesch B, et al (1985): A cellular gene encodes scrapie PrP_{27-30} protein. *Cell* 40:735–746

Prusiner SB (1982): Novel proteinaceous particles cause scrapie. *Science* 216:136–144

Rohwer RG (1984): Scrapie infectious agent is virus-like in size and susceptibility to inactivation. *Nature (London)* 308:658–662

Deafness

Joseph E. Hawkins, Jr.

Deafness (anacusis) is the complete inability to hear speech and other sounds, however much they may be amplified. Persons with partial deafness (hypoacusis, dyacusis) are correctly described as hard of hearing or as having a hearing loss. Chaucer was choosing his words carefully when he described his Wyf of Bathe as only "som-del deef."

The degree of hearing impairment is usually estimated by subjective audiometry, using pure tones or speech at precisely measured intensities to compare the patient's threshold of hearing with that of the normal listener. The difference, in decibels (dB), is the patient's hearing level: the greater the hearing level, the more severe the impairment. The audiogram shows this difference plotted as a function of frequency. Deafness has been defined as a hearing level of 92 dB ISO or worse (using the audiometric calibration of the International Standards Organization).

For some patients, objective audiometry is required. Electrocochleography records the electrical potentials of the cochlea and cochlear nerve in response to brief sounds; evoked response audiometry those of the brain stem and higher centers. A third objective technique measures the threshold for the acoustic reflex, the brief contraction of the stapedius muscle of the middle ear elicited by sudden, moderately intense sounds.

If congenital, hearing impairment may be the result of one of several types of genetic malformation of the ear, of disease in the mother during pregnancy (e.g., rubella, syphilis), or of perinatal disease or accident (e.g., neonatal jaundice, anoxia, trauma). If acquired, the impairment may be due to any of a variety of causes, including viral or bacterial diseases (e.g., mumps, herpes zoster oticus, meningitis, syphilis), head injury with temporal bone fracture, ototoxic drugs (especially certain antibiotics and diuretics, aspirin in large doses), exposure to intense noise, blast injury, inadequate blood supply to the ear caused by vascular insufficiency or occlusion, otosclerosis, Menière's disease, or simply the aging process itself, the commonest cause of all. Often both ears are equally affected, but in Menière's disease and in sudden deafness only one ear may be involved. In the latter condition, either occlusion of a small artery supplying the cochlea or a viral infection may be responsible.

The site of hearing impairment is usually peripheral, the result of defect or pathological change in the outer, middle, or inner ear. If the inner ear is normal and sound is prevented from reaching the cochlea by congenital malformation or by injury to the external canal, the drum membrane, or the ossicular chain of the middle ear, the resulting hearing loss is said to be conductive. Hearing for all frequencies tends to be affected, but the loss is not complete. Amplification of speech sounds with a hearing aid often gives a socially adequate improvement.

Otosclerosis most often takes the form of a conductive impairment, in which sound does not reach the cochlea because the stapes footplate is sealed in the oval window by abnormal deposits of bone. The operation of fenestration, producing a new oval window (fenestra) in the wall of the lateral semicircular canal, and stapes mobilization, freeing the footplate to vibrate again, represented the first successful attempts to restore hearing to otosclerotic patients by surgical means. Unfortunately, the improvement did not always last. Stapedectomy has now almost completely replaced older operations. The stapes is removed and a short length of wire or plastic attached to the long process of the incus is substituted, the oval window being covered with a bit of vein or fascia. Together with tympanoplasty for the correction of other middle ear defects, stapedectomy constitutes a dramatic advance in modern surgery.

In sensorineural hearing loss the middle ear is intact, but there are generally pathological changes in both the cochlea and the cochlear nerve. Most often, extensive loss of the ciliated sensory cells (hair cells) of the organ of Corti has occurred, with secondary degeneration of at least some of the cochlear nerve fibers and spiral ganglion cells. This is the type of hearing loss caused by aging, ototoxic drugs (Fig. 1), and noise exposure. The changes tend to occur first in the basal turn of the

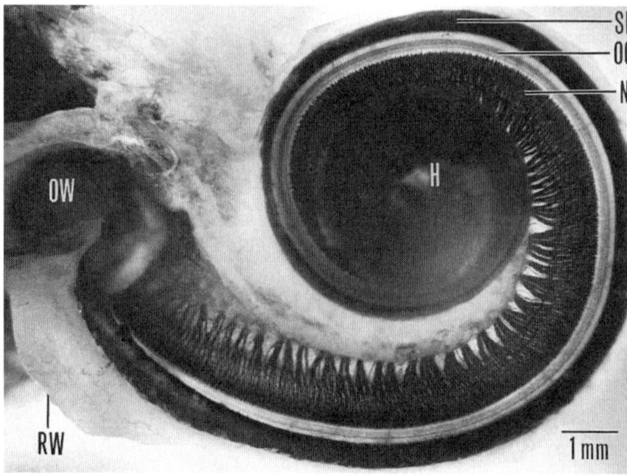

Figure 1. Normal-appearing right cochlea from a 25-year-old woman, prepared by microdissection after fixation with osmium tetroxide solution approximately 23 hours postmortem. OW, oval window, stapes removed; RW, site of round window; SL, spiral ligament; OC, organ of Corti, with rows of sensory and supporting cells seen as dark bands on translucent basilar membrane; N, myelinated cochlear nerve fibers in osseous spiral lamina; H, helicotrema, i.e., opening at apex between scala vestibuli and scala tympani. Reproduced by permission from Johnsson L-G, Hawkins JE (1972): Sensory and neural degeneration with aging, as seen in microdissections of the human inner ear. *Ann Otol Rhinol Laryngol* 81:179–193.

cochlea, which responds to the higher frequencies of sound, but they may spread to include the upper turns and thus affect the hearing for lower frequencies as well. A slow progression of this type of impairment is commonly observed in presbyacusis, the hearing loss that accompanies aging. During treatment with ototoxic drugs, especially with the aminoglycoside antibiotics and the loop diuretics in combination, a rapidly progressive hearing loss can occur. Similar changes occur in several inherited degenerative syndromes. Any such loss is permanent, since the sensory cells are not replaced (Fig. 2). The tissues of the cochlea may degenerate after viral diseases such as mumps, after severe prolonged Menière's disease, in cases of sudden deafness, and in the destructive capsular form of otosclerosis that is not limited to the stapes and oval window but violates the periosteal lining of the bony labyrinth. When the impairment is both conductive and sensorineural in character, it is described as mixed.

Because sensorineural loss tends to affect primarily hearing for higher frequencies, presbyacusic and other patients with this form of impairment have difficulty in understanding sibilant and fricative speech sounds and in distinguishing unvoiced stop consonants. They are at a particular disadvantage when trying to converse in noisy surroundings (the "cocktail party effect"). A hearing aid may be helpful for such a patient, but the results obtained are sometimes disappointing, because the patient may be unable to resolve the spectral details of speech sounds in the presence of noise, even after amplification.

Sensorineural impairment may be accompanied by disturbing symptoms such as tinnitus, a subjective ringing-in-the-ears that may also take the form of buzzing or roaring noise. Tinnitus often accompanies attacks of Menière's disease, in which the acuity of hearing tends to vary from time to time (fluctuant hearing loss). There may also be an abnormally rapid increase in the perceived loudness as the intensity level of sound increases: a phenomenon known as recruitment of loudness. Furthermore, unless the impairment of hearing is symmetrical, sound may produce unpleasant sensations of different pitch in the two ears (diplacusis).

The symptom of recruitment indicates a pattern of selective pathological change in the cochlea, with degeneration of many outer hair cells and preservation of most of the inner hair cells. When there is hearing loss and tinnitus without recruitment, a so-called retrocochlear lesion, such as an acoustic neurinoma or other tumor exerting pressure on the auditory nerve and brain system, may be present. Other types of retrocochlear lesions may be due to degenerative changes in the cochlear nuclei or other structures of the auditory pathways in the brain stem as a result of vascular accident or disease. In advanced age, the central processing and perception of speech may be affected by neuronal degeneration at higher levels of the auditory pathway.

Genetic forms of deafness and their underlying pathology can be studied in certain types of domestic and laboratory animals that are either born deaf or become deaf sometime after birth. Among these are the *waltzer, shaker,* and certain other strains of mice, blue-eyed white cats, and dalmatian dogs. The changes seen in the inner ear resemble closely those occurring in some types of congenitally deaf patients. Noise-induced and ototoxic hearing losses have been studied experimentally in guinea pigs, chinchillas, cats, and monkeys. Presbyacusic changes vary with the species examined. In the dog, they more nearly resemble those seen in aging human ears than do those in the largely vegetarian rhesus monkey.

In his *Journey to the Western Islands of Scotland,* Dr. Samuel Johnson, telling of his visit to one of the earliest schools for the deaf, describes deafness, with slight but pardonable exaggeration, as "one of the most desperate of human calamities." The rehabilitation of the deafened patient can be difficult, time-consuming, and expensive.

The deaf child presents an educational challenge of enormous magnitude and complexity. The deaf adolescent or adult often suffers from an isolation that the blind do not experience, so that psychiatric problems may constitute a serious additional handicap. The electronic cochlear prosthesis, implanted in the inner ear to bypass the defective end organ by delivering electrical stimuli representing speech and other sounds directly to the cochlear nerve fibers, offers hope for some deaf patients. It can give them an awareness of the acoustic environment as well as assistance in lip reading, even if it does not yet assure the understanding of speech. Continuing investigation will show for what types of deafness the cochlear prosthesis is likely to prove the greatest value.

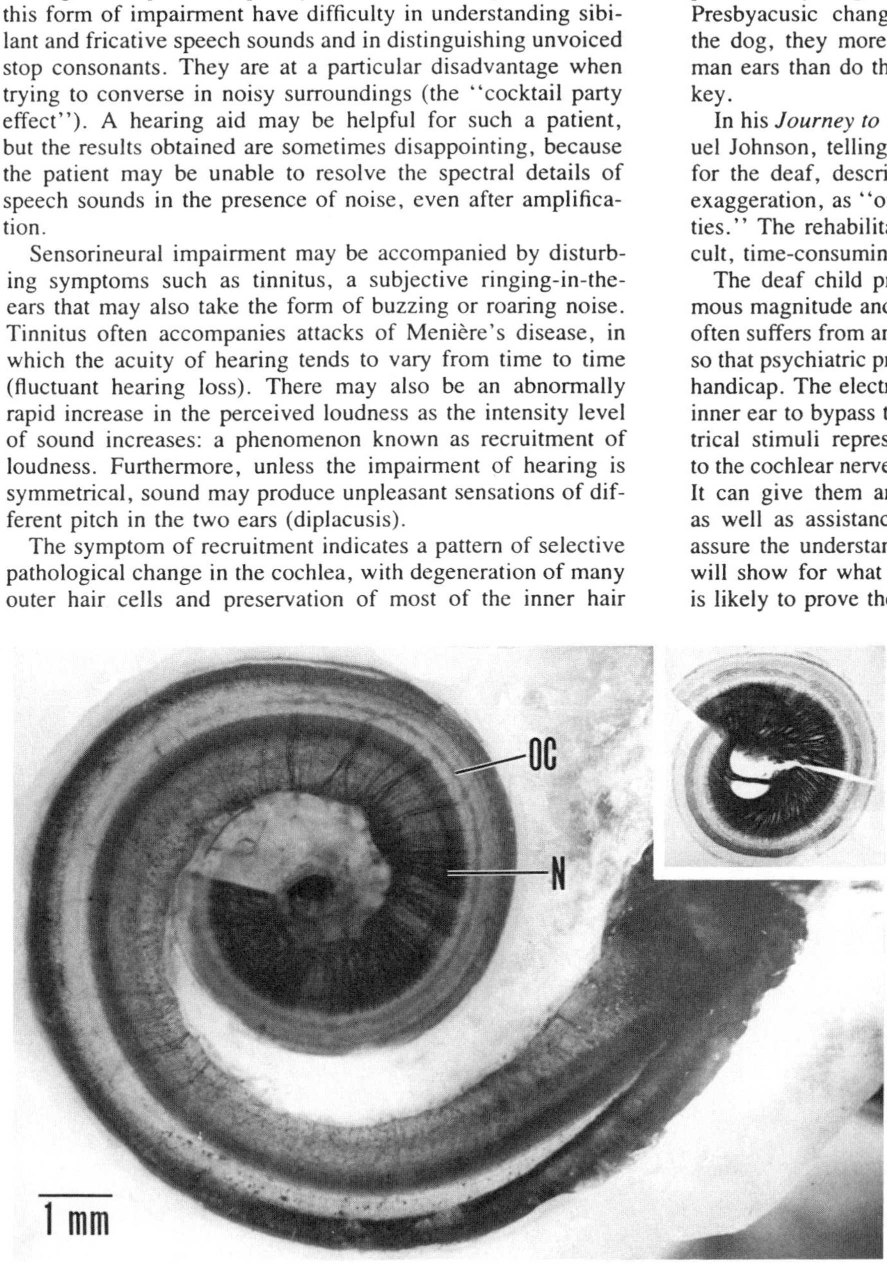

Figure 2. Left cochlea of a 51-year-old woman with ototoxic deafness caused by oral neomycin treatment over a two-year period. The organ of Corti has completely disappeared from the basal turn, and almost all the myelinated nerve fibers have degenerated. Supporting cells of Corti's organ, OC, and nerve fibers, N, are still present in the middle and apical turns (inset). Reproduced by permission from Johnsson L-G, Hawkins JE, Kingsley TC, Black FO, Matz GJ (1981): Aminoglycoside-induced inner ear pathology in man, as seen by microdissection. In: *Aminoglycoside Ototoxicity,* Lerner SA, Matz GJ, Hawkins JE, eds. Boston: Little, Brown.

Further reading

Ballantyne J (1984): *Deafness*, 4th ed. New York: Churchill Livingstone

Davis H, Silverman SR (1978): *Hearing and Deafness*, 4th ed. New York: Holt, Rinehart and Winston

Jerger J, ed (1984): *Hearing Disorders in Adults*. San Diego: College-Hill Press

Schuknecht HF (1974): *Pathology of the Ear*. Cambridge: Harvard University Press

Dementia

Fred Plum

Dementia is a sustained, multidimensional loss of cognitive function secondary to organic central nervous system damage, unaccompanied by evidence of an acute superimposed state of clouded consciousness as occurs with delirium or reduced arousal. The onset of dementia can be abrupt, maximal, and static, e.g., following cardiac arrest or severe head trauma, or progressive such as occurs in the degenerative diseases of aging. Of the two forms the progressive types of dementia create by far the more frequent problem and, reflecting the increased longevity of the population, have become one of the major public health concerns of the Western world. Over 6% of adults aged over 65 years and 20% over the age of 80 are estimated to suffer from a medically or socially disabling degree of dementia.

In most dementias, the clinical history discloses the subtle onset of disinterestedness, disorientation, personality changes, and diminished attention to self-care. Relatively trivial head trauma, intercurrent illness, or minor surgery such as cataract removal may precipitate more sudden deteriorations. Memory and nonspecific language abnormalities, particularly nominal amnesia, as well as emotional lability, depressive feelings, and sometimes, paranoid ideation are common features. A few patients suffer auditory and visual hallucinations. A large number of dementing illnesses, especially Alzheimer's disease, cause no early disturbances in motor or sensory functions. Others may produce extrapyramidal dysfunction, abnormalities of upper and lower motor neuron activities, urinary or fecal incontinence, and, rarely, seizures. Table 1 lists the principal causes of progressive dementia, and Figure 1 gives examples of some typical computerized tomography (CT) scans.

Most dementia reflects damage to the association cortex of the cerebral hemispheres or the subcortical thalamic nuclei that project to areas of association cortex. The neuropathology differs from disease to disease, in some producing predominant changes in neurons, in others appearing to choose both astrocytes and nerve cells as targets for maximal injury.

At least half of presenile dementia (appearing before age 65) and senile dementia cases are attributed to states in which the intellectual decline is accompanied by morphological changes in the brain of *Alzheimer's disease* (senile dementia Alzheimer type: SDAT). SDAT is predominantly a sporadic disorder, but appears to be linked to a genetic factor with approximately one-third reporting similar illnesses in blood relatives. An increased incidence of Down's syndrome (trisomy 21) and lymphoma also occurs in first-degree relatives. The condition affects females somewhat more frequently than males and, apart from changes in mental activity, usually spares other neurological functions early in its course.

The characteristic changes found postmortem in the brain of SDAT include moderate gross atrophy especially involving the frontal and temporal lobes. Microscopically, the most severe cell loss affects the large pyramidal cortical-cortical transmitting neurons in layers 3 and 5 as well as cells in the hippocampus and amygdala. Silver-staining plaques and intraneuronal fibrillary tangles, the hallmarks of the disease, distribute themselves most prominently in the same areas in numbers that roughly correlate with the degree of dementia. Immunohistochemical staining shows a prominent reduction in choline acetyltransferase (ChAT) in these regions, along with a marked reduction and disruption of somatostatin-staining neurons in layers 2, 3, and 5 as well as the hippocampus and amydgala. Substance P, but not other peptide transmitters examined to date, is also reduced in cortex. Cortical plaques contain an amyloid core plus assorted detritus that includes material staining for ChAT and somatostatin as well as fragments of other nerve cells and possibly glial material. Accompanying these changes is prominent neuronal loss in the basal forebrain nucleus of Meynert, the main source of cholinergic projections to the cortex, and, less consistently, in the monoaminergic locus ceruleus. These neuropathological abnormalities indicate a widespread degeneration of cortical-cortical connectivity in association and limbic areas in SDAT plus prominent loss of major nonspecific ascending cholinergic and monoaminergic autonomic systems. Presently, it is not possible to assign the primary pathologic change as arising either at cortical or subcortical levels, nor can one identify any single transmitter system that leads the brain into its decay. The cause of such widely diffused but, nevertheless, selective cell death remains unknown.

In clinical studies, CT or magnetic resonance (MR) imaging of the brain usually shows moderately advanced, diffuse cerebral atrophy marked by widened sulci and enlarged lateral ventricles. Metabolic studies of brain using positron emission tomography (PET) disclose a generalized reduction of oxidative metabolism usually most marked in parietotemporal areas of association cortex.

SDAT ordinarily runs its course in 5 to 10 years, although a few examples develop semiacutely and families occasionally describe patients in whom personality changes go back as far as 15 to 20 years before the appearance of fully developed dementia. Eventually, those affected become rigid and hypoki-

Table 1. Etiology of Progressive Dementia and Approximate Incidence (%)

Senile dementia of the Alzheimer type	50%
Multi-infarct dementia	10–15%
Mixed SDAT and MID	10–15%
Alcoholic-nutritional dementia	5–10%
Normal pressure hydrocephalus	5%
Miscellaneous: Huntington's disease, neoplasms, chronic subdural hematomas, Parkinson's disease, Creutzfeldt-Jakob disease, AIDS, and unknown cause	5–20%

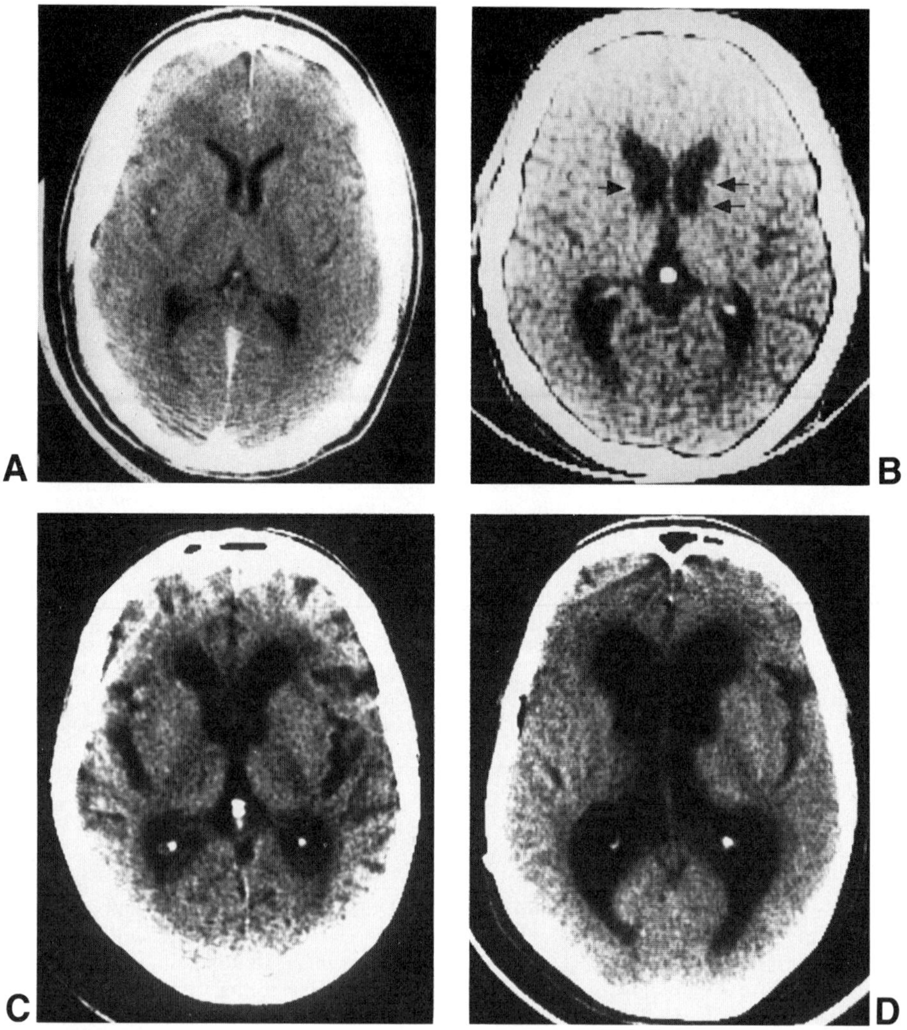

Figure 1. Computerized tomographic scans showing characteristic horizontal sections of brain in a normal person and in persons with dementia. A. Normal 53-year-old male. B. Early symptomatic Huntington's Disease in a 50-year-old male. Note caudate atrophy (arrows), more on right side of film than left. C. Primary alcoholic-nutritional atrophy in a 58-year-old male with Korsakoff's dementia. Note the enlargement of both the ventricular and subarachnoid spaces. D. Communicating hydrocephalus due to cryptococcosis in a 68-year-old demented woman. Note the considerable ventricular enlargement with relative obliteration of surface subarachnoid markings. A, C, and D, courtesy of Dr. Michael Deck, Department of Radiology, New York Hospital. B, courtesy of Dr. Stanley Fahn, Neurological Institute, College of Physicians and Surgeons of Columbia University, New York, NY 10032.

netic and in their terminal stages approach a totally vegetative state. No scientifically established treatment exists, although experimental efforts have attempted to induce benefit from the parenteral administration of cholinergic drugs.

Pick's disease, a much less common presenile dementia, runs an even slower course than SDAT. Small morphological differences distinguish the two conditions which may represent biological variants of a similar cause.

Intellectual loss due to successive strokes, so-called *multi-infarct dementia* (MID), accounts for approximately 10–15% of late-life mental decline. The disorder reflects the added effects of successive, multifocal widespread damage to cerebral areas regulating specific and nonspecific psychological functions. Mental changes commonly are accompanied by abnormalities in cerebral, motor, and sensory functions. As a rule, the vascular disorder pursues a progressive stepwise course, new episodes of minor or major deterioration reflecting additional vascular occlusions or thromboses and producing memory loss, aphasia, various agnosias, or sensorimotor defects. The neuropathology, as expected, consists of multiple infarcts, large and small, often accompanied by at least a moderately extensive amount of Alzheimer type cellular abnormalities.

Alcoholism and, to a lesser degree, substance abuse represent major causes of dementia. Exact figures of prevalence are difficult to construct since the linkage to brain damage is often indirect. Alcohol ingestion, for example, precedes and may precipitate as many as 40% of serious road traffic accidents, falls, and other head injuries and shows a high association with nutritional insufficiency as well as suicide attempts. Swedish investigators recently found that an estimated 40% of deaths among men aged 46–48 years were alcohol related, and in Finland, among men less than 40 years old with acute stroke, at least 20% suffered their attack shortly after an episode of heavy drinking. A Danish study has reported that 28 of 37 male heavy drinkers under the age of 35 years suffered subjective symptoms of mental impairment, and almost one-fifth tested as being frankly demented using standard appraisals of intelligence capacity.

The mechanisms of alcoholic dementia are only partly known; many undoubtedly are nutritional but some must be directly toxic as well. Among alcoholic men under 40 years, about half show cerebral atrophy by CT scan, and the incidence rises as age advances. The structural basis of these changes is not well delineated although some evidence indicates that they may be partly reversible in their early form. Brain atrophy and the specific abnormalities of Wernicke's disease and Korsakoff's psychosis are almost routine among chronic, heavy alcoholics whether or not they die from the direct effects of the disease and its immediate complications.

Chronic communicating hydrocephalus, a condition sometimes called *low-pressure hydrocephalus* because cerebrospinal fluid (CSF) pressure usually falls within or only slightly above the normal range, causes or contributes to approximately 5% of late-life dementias. The disorder is more frequent in men than women and most often probably results from meningeal adhesions or scarring arising long after the everyday head trauma of younger life or an unremembered meningeal infection. In at least half the cases no antecedent cause can be found. The mechanism of the disorder is believed to reflect impairment of CSF passage over the surface of the brain toward its absorption points in the pacchionian granulations of the large venous sinuses. In some instances, passage through the granulations themselves may be impeded. Areas of the subarachnoid spaces become plastered down, and the cerebral ventricles enlarge with most of the brain atrophy in the early stages of the disease involving the periventricular areas of the hemispheres. Diagnosis rests largely on a characteristic clinical history, accompanied by the appearance on CT or MR scans of widely dilated cerebral ventricles. Concurrently, the images show relatively little brain atrophy over the surface, especially at the cranial vertex. Affected patients develop an insidiously beginning and slowly progressing dull, apathetic intellectual loss often accompanied by a broad-based, ataxic gait and intermittent sphincter incontinence. Neurosurgical shunting of the CSF from the lateral ventricle into the heart or peritoneal cavity provides at least some improvement in about half the cases carefully chosen according to these criteria. A variety of procedures have been evaluated in unsuccessful efforts to improve the prediction of who will do well following surgery.

Among less frequent structural causes of dementia, those associated with *Parkinson's disease* (PD), *Huntington's disease* (HD), and *Creutzfeldt-Jakob disease* (CJD) possess particular biological interest.

About 25% of patients with PD undergo a mental decline sufficiently severe to prevent them from remaining completely independent. Clinically, the intellectual changes usually begin insidiously and progress slowly with symptoms resembling those of SDAT. Neuropathologic alterations in the brain also resemble those of SDAT and include plaques and tangles in the cortical gray matter along with reduction of ChAT in the cortex and basal nucleus of Meynert (NBM). Some workers have found a somewhat lesser reduction in cortical ChAT and greater reduction in NBM neuronal loss in PD than in SDAT. The finding and its possible significance awaits confirmation and explanation.

HD has attracted considerable recent attention because of a substantial increase in the understanding of the disorder's neurobiology and genetics. HD is a progressive, eventually fatal dementia transmitted as an autosomal dominant trait with complete penetrance. Clinically, the disorder appears most often during the third to fifth decade of life. Mental decline, often accompanied by prominent emotional and personality changes, usually heralds the onset, accompanied or shortly followed by generalized dance-like, quick choreiform as well as slower dystonic movements. Medium-sized spiny GABAergic cells in the striatum undergo a prominent and early reduction in number; several other classical neurotransmitters and neuropeptides show less prominent changes (Table 2). Additional cell loss involves the ventral thalamic nucleus and subthalamus, and inconsistently located neuronal dropout affects the cerebral cortex, most markedly in the frontal lobes. The brain as a whole shrinks by about 30% by the time death supervenes. Recent molecular genetic studies have identified a restriction fragment length polymorphism mapped to chromosome 4 that is linked to the HD gene. The anomaly has been identified in two large, geographically separated kindreds with HD, one located on Maracaibo Bay in Venezuela, the other in Iowa. The finding promises to provide in the future accurate presymptomatic and prenatal diagnosis of carriers and may pave the way to effective prevention or treatment once the specific molecular nature of the genetic error is identified.

CJD is a rare, rapidly developing and progressive dementia affecting persons of either sex in their middle and late middle years. Clinically, the disease produces an afebrile, progressive intellectual decline of a multidimensional character combined usually with the development of prominent signs of corticospinal and basal ganglia dysfunction. Sooner or later, focal or multifocal myoclonic seizures develop. Pathologically, one finds in the brain a moderate neuronal loss accompanied by extensive astrocyte proliferation and swelling, the latter giving a spongiform appearance to histological sections examined by light microscopy. CJD mainly occurs sporadically but also emerges in small family clusters. Its cause lies in an as yet incompletely understood or identified, nonimmunogenic infectious agent that can be transmitted via cerebral inoculation into a number of animal species, including various primates and small rodents. Several months to a year or more separate the time of inoculation from the development of morphological or behavioral signs of infection in the recipient; this feature, plus neuropathological similarities to visna in sheep and kuru in the Fore tribe of New Guinea place the disease among the slow virus infections of humans. Contagiousness by the usual person-to-person contact is extremely low, indeed probably negligible in practical terms, but human passage has occurred from postmortem corneal transplants taken from unrecognized

Table 2. Neurotransmitter Changes in HD

Neurotransmitters decreased in HD	Neurotransmitters unchanged in HD	Neurotransmitters increased in HD
1. GABA (and GAD	1. Dopamine	1. Dopamine[a]
2. Acetylcholine (and CAT)	2. Serotonin	2. Thyrotropin-releasing hormone
3. Substance P	3. Vasoactive intestinal polypeptide	3. Somatostatin
4. Enkephalins	4. Norepinephrine	4. Neurotensin
5. Cholecystokinin		

From Martin (1984).
[a] Reported to be increased in some studies, but not in others.

cases. Sterilization of the agent is extremely difficult, and person-to-person transmission has followed the reuse of intracerebral electrodes implanted to study seizures in persons later diagnosed as having CJD.

A small number of structural, metabolic, and infectious disorders can produce a sustained decline in mental function resembling progressive dementia in their longevity but potentially reversible if detected and treated sufficiently early. Included in this category are selected cases of cryptic subdural hematoma or brain tumor, Wilson's disease, chronic liver cirrhosis, cyanocobalmine (B_{12}) deficiency, chronic sedative ingestion, thyroid, parathyroid, or adrenal endocrinopathies, and chronic central nervous system infectious granulomatous diseases such as syphilis, sarcoid, cryptococcus and the like. These so-called treatable dementias are relatively easy to identify by routinely employed tests, and substantial improvement often follows appropriate therapy.

Late-life nonorganic psychological depression often produces a condition called *pseudodementia* that sometimes can be difficult to differentiate from organic dementia. Depressed patients often become apathetic, quiet to the point of muteness, and demonstrate a pronounced loss of interest in their surroundings, sometimes accompanied by considerable anxiety. All these symptoms can also mark the early or intermediate stages of dementia. Where the diseases differ lies in the often prominent and repeated complaints by depressed patients that they are losing their mind, unable to concentrate, unable to give attention to things, and feel their strengths seeping away with time. On examination, depressed patients may not always respond to tests of mental function. When they do, however, they answer most of the standard questions correctly. By contrast, demented patients may be slow, but they usually respond fairly consistently, predictably getting the answers wrong. Furthermore, psychomotor retardation and absolute insomnia far more often mark the course of depression than dementia. Demented patients often develop a disordered sleep pattern but their total sleep hours decline relatively little and may even increase. Accurate distinction between dementia and depressive pseudodementia is not always easy, especially when the early dement retains enough insight to become depressed over his failing powers. Regrettably, neither biological markers nor consistent clinical features will measure accurately the degree of reversible psychological dysfunction in such instances. In many cases, the physician can only treat the psychological symptoms and wait, knowing that antidepressant treatment often reduces suffering and that time strengthens diagnostic accuracy in the clinical management of most dementias.

Further reading

Cummings JL, Benson DF (1983): *Dementia, A Clinical Approach.* Boston: Butterworths

Martin JB (1984): Huntington's disease: new approaches to an old problem. *Neurology* 34:1059–72

Morrison JG, Rogers J, Scherr S, Benoit R, Bloom FE (1985): Somatostatin immunoreactivity in neuritic plaques of Alzheimer's patients. *Nature* 314:90–92

Roberts GW, Crow TJ, Polak JM (1985): Location of neuronal tangles in somatostatin neurons in Alzheimer's disease. *Nature* 314:92–94

Terry RD, Katzman R (1983): Senile dementia of the Alzheimer type. *Ann Neurol* 14:497–506

Down Syndrome

Charles J. Epstein

Down syndrome is the set of physical, mental, and functional abnormalities that result from trisomy 21, the presence in the genome of three rather than the normal two chromosomes 21. The physical abnormalities that together give rise to the distinctive facial appearance associated with this condition include upslanting palpebral fissures with inner epicanthic folds, flatness of the bridge of the nose, midfacial hypoplasia, and a tendency to protrude the tongue, especially when very young. Many other functionally inconsequential minor abnormalities of the ears, hands, and feet may also be present, and stature is generally reduced. Approximately 40% of affected individuals are born with congenital heart disease, with endocardial cushion and related septal defects frequently being present. Obstruction of the intestinal tract also occasionally occurs during development. Although trisomy 21 is the autosomal trisomy most compatible with survival through the period of gestation, only about a third of affected embryos and fetuses are actually liveborn.

Individuals with Down syndrome are more than normally susceptible to infections, probably because of abnormalities of the T lymphocyte system. They also have an increased although still low likelihood of developing childhood leukemia. Despite these predispositions, their mean life expectancy, if they survive the first year of life, is now between 55 and 60 years.

Down syndrome is characterized by three salient clinical abnormalities of the nervous system. The first, recognizable immediately after birth and prominent in the first years of life, is a significant degree of hypotonia that produces a sense of floppiness or looseness. The second is mental retardation, which is generally moderate but may vary from mild to severe. The rate of acquisition of skills and capabilities is often normal during the first few months of life. However, there is then a decrease in rate so that the development of more advanced physical and intellectual skills such as the ability to construct sentences, to walk stairs alone, and to dress may be delayed by a year or more. Persons with Down syndrome have learned complex intellectual tasks such as reading, writing, and arithmetic calculations, but this does not occur in all instances. There is evidence that the intellectual achievements of affected individuals reared in enriched environments, such as the home, are greater than was previously observed in trisomics confined to institutions for the mentally retarded. However, although a wide variety of agents have been tried, including megavitamins and 5-hydroxytryptophan, there is not, at present, any accepted pharmacologic treatment of Down syndrome.

The exact anatomical and physiological causes of the hypotonia and mental retardation in Down syndrome are unknown. Very modest decreases in head circumference and brain weight appear to be present, with the cerebellum and brain stem being particularly affected. Although many anatomical and histological anomalies of the nervous system have been described, including heterotopias and abnormalities of dendritic spines, most appear to be nonspecific and inconsistent, and gross structural abnormalities cannot explain the neurological deficits. However, recent studies have indicated that there may be, early in life, selective decreases in the density of neurons in particular layers of the cortex, and possibly decreased numbers of synaptic contacts as well. It has been suggested that such findings may reflect maturational delay rather than frankly abnormal neuronal differentiation and migration in fetal and infant trisomic brains. Neuronal deficits have also been observed in young adults, but it is not clear whether these represent developmental or degenerative changes.

The possibility that metabolic and physiological, as opposed to structural, abnormalities may contribute to the neurological dysfunction in Down syndrome has been widely considered, but at present there is no convincing evidence that any specific metabolic aberration has a significant role in the pathogenesis of this condition. Considerable interest has focused on peripheral neurotransmitter function in Down syndrome, and hypersensitivity to the mydriatic effect of atropine has been well documented. However, there are no data in support of central neurotransmitter abnormalities in young individuals with trisomy 21. Some data suggestive of electrophysiological aberrations do exist. These include abnormal patterns of visual and auditory evoked potentials and abnormal electrical properties of cultured fetal trisomic dorsal root ganglion neurons.

The third clinical abnormality of the nervous system, recognizable in 25 to 50% of individuals with Down syndrome living into the fifth decade, is the progressive loss of mental function resulting from the superimposition of a dementing process on the preexisting mental retardation. Associated with the development of these clinical features of dementia has been the appearance, in virtually all Down syndrome individuals over 35–40 years of age, of the pathological and neurochemical changes associated with Alzheimer disease. Brains of adults with Down syndrome possess granulovacuolar cytoplasmic changes, senile plaques, neurofibrillary tangles, and cerebrovascular amyloid indistinguishable from those found in the brains of individuals with Alzheimer disease. Chemical similarities, including decreases of cholinergic and other neurochemical markers, have also been noted. Persons with Down syndrome thus constitute a unique population at very high risk of developing Alzheimer disease, both pathologically and, in many instances, clinically. However, the nature of the association between the two conditions is unknown. While Alzheimer disease may be a direct consequence of genetic imbalance produced by trisomy 21, it is more likely that the trisomic condition serves to make individuals with Down syndrome more susceptible to those factors that lead to the development of Alzheimer disease in a chromosomally normal aging population.

The pathogenetic relationship between the presence of an extra chromosome 21, which is presumed to result in a 50% increase in the expression of each of the active genes on this chromosome, and the several phenotypic features of Down syndrome described above is unknown. Current research is directed toward identification of these genes and to assessing the consequences of their overactivity. Seven chromosome 21 genes of known function have so far been recognized, and knowledge of these genes has in turn been used to develop an animal model for Down syndrome. It has been shown, by comparative gene mapping, that four of the known human chromosome 21 genes are present on mouse chromosome 16, and it has been inferred that many other chromosome 21 genes are present on mouse chromosome 16 as well. Mouse trisomy 16 thus constitutes a genetic model of human trisomy 21 that is suitable for in vivo and in vitro developmental, structural, and chemical studies. It is hoped that these and other approaches will reveal not only how trisomy 21 results in the abnormalities associated with Down syndrome, but also whether anything can be done therapeutically to improve the mental and physical development of affected individuals.

Further reading

Epstein CJ (1986): *The Consequences of Chromosome Imbalance: Principles, Mechanisms, and Models*. New York: Cambridge University Press

Pueschel SM, Rynders JE, eds (1982): *Down Syndrome: Advances in Biomedicine and the Behavioral Sciences*. Cambridge: Ware Press

Scott BS, Becker LE, Petit TL (1983): Neurobiology of Down's syndrome. *Prog Neurobiol* 21:199–237

Dyslexia

Albert M. Galaburda

In the United States dyslexia refers only to developmental disorders of reading. In the United Kingdom and parts of Europe the term dyslexia is used to refer to a variety of reading difficulties, which include both developmental and acquired disorders. Dyslexia is considered here as a reading disorder of developmental origin.

In the absence of well-characterized biological markers, dyslexia is defined operationally as "a failure [by children] to learn to read with normal proficiency despite conventional [or even specialized] instruction, a culturally adequate home, proper motivation, intact senses, normal intelligence and freedom from gross neurologic deficit" (Eisenberg, 1967). The components of this definition, i.e., normal proficiency, adequate instruction and proper motivation, a tolerable degree of visual and auditory impairment for the purpose of learning to read, the acceptable range of normal intelligence, etc., must be further specified, and this turns out to be a major source of disagreement among authorities. However, it remains that a small but significant proportion of school-age children fail in their attempts to learn to read despite excellent instruction and apparent intellectual ability.

The exact characterization of the elements making up the definition of dyslexia will determine the figures on the prevalence of the condition in the population. Thus, for example, a mere change of 5 points in the IQ range will include or exclude large numbers of children. For this reason prevalence figures in the literature vary between 2% and 16% of school-age children. In studies carried out in the Isle of Wight, reading retardation defined as more than 2 standard errors below expected reading level in children with a mean IQ of 102.5 was present in between 3.5% and 6% of the children. The same studies found a male-to-female ratio of 3.3 to 1; this male predominance has been noted by most other studies.

Inheritance plays an important role in the transmission of dyslexia. There were early suggestions of a single autosomal dominant gene, whereby male-female differences were explained on the basis of unequal societal interest in the search for and diagnosis of dyslexia among boys and girls. More recently sex differences in dyslexia have been explained by the apparent differences between normal males and females in the brain organization for language. In that case genetic disturbances occurring equally in boys and girls would be apt to produce different effects. The most recent hypothesis argues for specific influences of sex hormones *in utero* in the expression of faulty gene(s). Abnormal testosterone activity is implicated, explaining the increased prevalence in males.

The first reference to difficulty with the written word in two otherwise normal subjects appeared in Eugene Labiche's *La Grammaire* (1867). Subsequently, in the United Kingdom, the first medical mention of specific reading disability (dyslexia) by W. Pringle Morgan, and the first serious attempt to discuss dyslexia in the light of contemporary neurological knowledge by James Hinshelwood. Soon thereafter additional reports came out of the European continent and the United States. Hinshelwood considered the disorder to reflect problems in brain development; Samuel Orton in the United States emphasized the relationship between dyslexia and disordered brain asymmetry, and Hallgren in Denmark underscored the familial nature of the disability. Research in recent years has stressed the cognitive features and classification of dyslexia as an aid to diagnosis and as a guide to the development of educational methods of remediation.

Dyslexics represent a heterogenous group with regards to cognitive-linguistic-neurological characteristics. At least one common subtype exhibits disordered linguistic processes. There is another group of children in whom language deficits are difficult to demonstrate in the face of severe visuospatial and visuomotor anomalies. Despite the significant extralinguistic abnormalities in this latter group, it is not possible to state that they are etiological in the reading disorder, and may simply accompany a more generalized disturbance also affecting language. A third group appear to show features of the first two. Furthermore there seem to be compensatory linguistic and developmental neural strategies available to some individuals that modify the behavioral expression in unpredictable ways and produce still additional smaller subgroups.

A variety of neurological accompaniments are associated with dyslexia. These range from so-called minor neurological signs through changes in standard patterns of brain asymmetry on computerized radiographic scans and changes in brain electrical activity to alterations in brain structure. Some of the neurological signs relate to malfunction of cognitive processes known from adult brain injury cases to be associated to specific brain regions; others may include skeletal anomalies such as scoliosis, motor problems involving dexterity of one or both hands and the extraocular muscles, and stuttering. Electroencephalographic abnormalities range from focal slowing and spikes to frank epileptiform activity.

Paralleling reports of increased non-right-handedness in dyslexia and alterations in the standard patterns of structural asymmetry of the brain have also been noted. Approximately two-thirds of unselected brains show language-related regions in the left hemisphere to be larger than comparable regions on the right; 10% show the reverse pattern and about 25% of the brains are symmetrical in these regions. It appears that dyslexics show aberrant patterns more often, and frank reversal of asymmetry commonly, especially among dyslexics with low verbal IQ. Five brains of dyslexic individuals, all males, have been studied after death. The ages of the subjects ranged from 12 years to 30 years. In four of these brains the pattern of asymmetry was studied and fell into the symmetrical category, a finding that is statistically significant.

All five brains of dyslexics studied for anatomical changes have shown developmental cortical anomalies. These are of two types: ectopias and dysplasias. Ectopias (the presence of structures in places from which they are normally absent) consist of collections of neurons and abnormal tufts of fibers and blood vessels predominantly in the molecular layer of the cortex. Often frank verrucous excrescences are seen. Dysplasias consist of disordered lamination of cortex and cell-free areas, sometimes forming frank micropolygyria. The anomalies are present predominantly in the inferior frontal gyrus and on the superior temporal gyrus and temporoparietal junction, and involve the cortex of the left hemisphere in great excess over that of the right. No brains free of anomaly have been reported. It has been suggested that pathological effects *in utero* preferentially affect the development of the left hemisphere and its language areas, thus resulting in linguistic anomalies that show up most clearly in the reading process.

The presence of brain changes in dyslexia has prompted the search for additional biological associations. It has been found that there is an association between left-handedness, learning disability (including dyslexia), and immune disease. The immune disorders primarily involve the bowel and thyroid, although additional conditions are still under study. It has been suggested that the same hormonal influences in fetal life responsible for altered left hemisphere development and dyslexia account for the excess of left-handedness and immune-related disorders, the latter through testosterone effects on the development of the thymus and other immune processes. Furthermore selected strains of mice that develop similar immune-based disease frequently exhibit alterations in cortical development consisting of neuronal ectopias and dysplasias.

Further reading

Reviews and symposia

Eisenberg L (1979): Definitions of dyslexia: Their consequences for research policy. In: *Dyslexia. An Appraisal of Current Knowledge*, Benton AL, Pearl D, eds. New York: Oxford University Press

Hynd G, Cohen M (1983): *Dyslexia*. New York: Grune & Stratton

Anatomical findings

Galaburda AM (1983): Developmental dyslexia: Current anatomical research. *Ann Dyslexia* 33:41–53

Galaburda AM, Kemper TL (1979): Cytoarchitectonic abnormalities in developmental dyslexia: A case study. *Ann Neurol* 6:94–100

Galaburda AM, Sherman GF, Rosen GD, et al (1985): Developmental dyslexia: Four consecutive patients with cortical anomalies. *Ann Neurol* 18:222–233

Biological associations

Geschwind N, Behan P (1982): Left handedness: Association with immune disease, migraine, and developmental learning disorder. *Proc Natl Acad Sci USA* 79:5097–5100

Eating Disorders

Domeena C. Renshaw

In the wild, land mammals are remarkably similar in their seasonal accumulation of body fat. They are rarely obese. Domestic or captive animals, and active or sedentary humans, however, show much variation in fat deposits. Low activity and excess food intake appear to be factors in fattening agricultural and domestic animals. Farming provides protection from predators and no dominance challenge, preservation (even severe restriction) of territorial boundaries as in poultry and cattle coops, and controlled breeding rather than natural selection. Furthermore, synthetic hormones are used to enhance livestock weight. These additives may have unknown health and weight-gain consequences in humans.

The primitive task of hunting and food gathering for each meal has evolved for humans into money gathering, then grocery shopping, and storage of the fresh or dry edibles. Scientifically enhanced food production, 24-hour diners, and fast food outlets have made possible ingestion of the abundance available at any time outside or inside the home, thanks to freezers, refrigerators, canning, and drying. Affluence, novelty, and aggressive advertising have greatly added to the attraction of eating light, tasty, high-caloric snacks far beyond each person's natural hunger-satiation rhythms. In times of plenty very few children or adults learn to tolerate hunger feelings. Overfulness may be associated with a sense of emotional well-being, so that overeating can become customary, resulting in overweight.

It may be argued that for primitive, nomadic tribes, stored energy as fat had survival advantages in times of food shortage. Were these people leaner or heavier than today? It is not known. Stone-age clay sculptures depict obese persons way before agriculture, but these artifacts may have been of exceptional rather than average citizens. Modern men and women do not claim (as in earlier centuries) that overweight is a protection from future hunger, tuberculosis, or other diseases. Periodic fasting (as in Lent or Yom Kippur) may be externally imposed by a belief system and is still widely practiced. Fasting chosen as a lifestyle leading to holiness is considered ascetic in many religions and cultures. Hunger strikes evolved as a form of temporary political protest and became well known at the turn of this century when women in England struggled for equal rights.

Fasting combined with nonviolence subsequently became a powerful political means to achieve change when used by Ghandi in Africa and in India. Fasting in this sense is not regarded medically as an eating disorder.

It remains controversial whether obesity (excess adipose tissue) is regarded as a disease entity. It is not listed in DSM I, II, or III (editions 1–3 of the American Psychiatric Association's *Diagnostic and Statistical Manual of Mental Diseases*) as an eating disorder with psychogenic components. Yet, undisputedly, it is still the most common disorder of metabolism, and the emotional precursors and sequelae are numerous. Diseases closely associated with obesity include cardiovascular diseases, hypertension, adult-onset diabetes, cholecystitis, infertility, arthritis, thromboembolism, varicose veins, and hernias. Sexual problems caused by low self-esteem, partner criticism, and mechanical or mobility difficulties may also be associated with obesity. Diets and how to lose weight are constant questions brought to physicians by those who are truly obese and those fearing it.

A common, although arbitrary definition used for obesity is 20% above the mean ideal body weight for men and women, corrected for height and age, usually taken from life insurance tables. Estimates are that in the United States 20% of men and 40% of women over 40 are obese by this measure.

Eating disorders have received media attention in the past 20 years. Headlines and articles about "the starving disease" or the "thin disease" (anorexia nervosa) in the 1970s were followed in the 1980s by reports of the unexpectedly high incidence of self-induced vomiting (bulimia) among college girls to keep their body weight at 15% to 25% below norm. These two sister eating disorders—anorexia nervosa and bulimia nervosa—are inextricably bound to obesity. DSM-III diagnostic criteria for anorexia nervosa are as follows:

Refusal to maintain body weight over a minimal normal weight for age and height.
Weight loss of at least 25% of original body weight.
Disturbance of body image with inability to perceive body size accurately.
Intense fear of becoming obese. This fear does not diminish as weight loss progresses.
No known medical illness to account for weight loss.
Amenorrhea.

Diagnostic criteria for bulimia are as follows:

Episodic binge eating pattern usually in less than 2 hours accompanied by an awareness of the disordered eating pattern with a fear of not voluntarily being able to stop eating; and depressive moods and self-depreciating thoughts after binging.

At least three of the following:

Rapid consumption of food during a binge;
Consumption of high-calorie, easily ingested food during a binge;
Inconspicuous eating during a binge;
Terminating binge eating by abdominal pain, sleep, social interruption, vomiting, or self-induced vomiting;
Repeated attempts to lose weight by severely restrictive diets or self-induced vomiting;
Eating pattern of alternate binges and fasts;
Use of laxatives for weight control.

All three groups—obese, bulimics, anorexics—have an internally driven compulsive preoccupation with food and verbalize dislike of being or becoming fat. Many girls may as a group of plump 12-to-14-year-olds begin to diet. Most do not persist, but a few will keep on dieting into an anorexia or bulimia nervosa state while others will persist with overeating into lifelong obesity. In the 1980s more young women seem to choose the bulimic pathway. Obese persons will give up on a diet and continue to eat despite a stated negative attitude toward being overweight. Anorexics and bulimics, however, will develop such an intense morbid abhorrence of obesity that their paths diverge. Political fasting is done by average-weight persons from abhorrence of perceived social or moral evils. Religious fasting is done from fear of breaking taboos or in search of higher good through self-mortification. Anorexics abhor fat, imperfection, and indulgence. Bulimics cycle a feast-vomit-fast pattern, although they also abhor fat and imperfection in themselves. Suicide potential is of concern due to their mood swings, which may respond to medication. So saint and suffragette, actress, fashion followers, models, mystics, and zealots have shared fasting behaviors with very different underlying motives, and often with much more complexity than any classification of eating disorders can provide.

For all three groups—obese, bulimics, anorexics—the psychological level of distress may be equally high regarding actual body weight and the affected individual's internally perceived body image. Self-blame, shame, regret, remorse, and guilt are accompanied by insecurity, anxiety, depression, and low self-esteem. Difficulty dealing with anger and suicidal thoughts are constant concomitants for all three groups.

All appetites, whether for food, for sleep, or for sex, are, in a complex way, on a somatopsychic interface where both physiology and feelings are affected when problems occur. Appetites are all vital functions. Only the control of sexual expression is not harmful to an individual (although that choice may destroy a relationship). Therefore, in the treatment and research of eating disorders, sexuality is hardly mentioned due to concern about health and survival. Exactly how the self-defeating compulsive, repetitive, often self-destructive, distressing behavior of eating disorders is neurophysiologically driven by each individual's emotions is still speculative. Interesting to note is that although alcohol is not essential to life but food is, the concept of "foodaholic" has been readily accepted due to the obese person's acceptance of job and family consequences of the excess weight plus secret and "cheat" eating. Overeaters Anonymous has adopted and adapted the Alcoholics Anonymous group support system. So to some extent have bulimics in their BASH (bulimia anorexia self-help) organization. Anorexics, however, seem to me to represent an inverse addiction in that they may lie, cheat, and pretend that they ate when they did not. Clinical management is difficult in all these conditions since no person can assume 24-hour guardianship for what another person eats. Psychosocial management must proceed, but the search for biological answers is also important. Theories of disturbed norepinephrine, amines, peptides, and neuroendocrine and hypothalamic function are being studied in obesity, bulimia, and anorexia, thus far inconclusively. Could there be special limbic system-hypothalamic pathways that await identification and understanding in all the eating disorders? And are there absent or even excess connections to the sexual centers of the pituitary-hypothalamic areas?

Further reading

Bierman EL (1982): Obesity. In: *Cecil Textbook of Medicine*. Wyngaarden JB, Smith LH, eds. Philadelphia: WB Saunders

Halmi KA, Falk JR, Schultz E (1981): Binge-eating and vomiting: A survey of a college population. *Psychol Med* 11:707–711

O'Neill CB (1982): *Starving for Attention*. New York: Continuum

Palmer RL (1980): *Anorexia Nervosa: A Guide for Sufferers and Their Families*. New York: Penguin Books

Pope HG, Hudson JI (1984): *New Hope for Binge Eaters*. New York: Harper & Row

Pyle RL, Mitchell JE (1983): The bulimia syndrome. *Female Patient* 8:48–53

Renshaw DC (1982): Obesity and sexual problems. *Female Patient* 7:58–60

Renshaw DC (1982): Sexual anorexia nervosa? *Chicago Med* 85(11):590–592

Epilepsy

Massimo Avoli and Pierre Gloor

Epilepsy is a disorder of brain function characterized by the episodic recurrence of paroxysmal neurological or behavioral manifestations caused by abnormal synchronous and excessive discharges of large groups of neurons. The ancient Greek word *epilepsia* literally means "seizure" and the disease itself was probably recognized prior to the development of the earliest civilizations.

Clinical features of epilepsy

The incidence of human epilepsy is estimated to range between 0.5 and 2% of the general population. Epilepsy is not a unitary disease. It has no single cause; rather it is a multifactorial condition reflecting acquired and genetic factors. Among the former can be listed perinatal and postnatal cerebral trauma, infections of the central nervous system (CNS), brain tumors, cerebral vascular lesions, congenital malformations, and some metabolic disorders. A period of months to years can elapse between the impact of the exogenous causal event and the appearance of the clinical symptoms. Knowledge of the exogenous factors involved in the causation of epilepsy is relevant for its prevention, since the incidence of the disorder can be reduced by improved perinatal care or by the prevention of brain injuries. Genetic factors include seizures caused by a single gene disorder (e.g., some inborn errors of metabolism), and others that to a greater or lesser degree interact with exogenous factors. In some forms of inherited seizure disorders in animals great progress has recently been made in understanding their molecular biology. No similar progress has yet been made for human epilepsy.

Epileptic discharge may involve any structure of the CNS although it can originate in only some of them (most often the cerebral cortex, including the hippocampus, or the amygdala). Consequently, though its most dramatic clinical manifestation is loss of consciousness with generalized tonic-clonic convulsions (which may be the final outcome of any epileptic seizure), many seizures exhibit a great variety of clinical signs and symptoms. Their distinctive features depend upon differences in the site of origin and in the extent and pattern of spread of the seizure discharge. Thus, the salient symptomatology of an epileptic seizure reflects the functional significance of the brain area involved in the seizure discharge. For example, twitching of half the face with discharge involving the contralateral face area of the precentral gyrus, phosphenes in half the visual field with discharge in the contralateral visual cortex, evocation of past memories in temporal lobe seizure discharge, etc., result from ictal activation of neuronal mechanisms represented in these areas, while other manifestations like the inability to utter or understand speech result from disruption of the normal function of speech cortex when it partakes in ictal discharge.

Depending on whether only a limited part of or the entire brain is involved in the seizure discharge at its onset, as judged by both clinical and electroencephalographic (EEG) criteria, epileptic seizures are subdivided into two main categories: (1) partial (i.e., focal) and (2) generalized. Each category is further subdivided into subcategories depending on the ictal symptomatology of the attack. In the differential diagnosis of epilepsy, it is necessary to consider other episodic disturbances of CNS function (e.g., syncope, migraine, hysteria, and other psychiatric disorders) and to determine whether the attacks are symptomatic of a structural brain lesion or a metabolic derangement or are mainly due to a genetically determined brain disorder. In many cases the etiology remains unknown.

The EEG examination, which detects signs of epileptogenic neuronal dysfunction, is the most important laboratory test used in the diagnosis of epilepsy. The practical value of the *EEG* rests on its ability to detect evidence for such a dysfunction even between epileptic attacks. The EEG signs of epileptic discharge can therefore be divided into (1) interictal discharges, present between seizures, and (2) ictal discharges, recorded only during an attack. Interictal discharges are characterized by short-lasting EEG abnormalities, such as high-amplitude spikes or sharp waves (Fig. 1A), while ictal discharges are more varied, but usually consist of an abnormally rhythmic

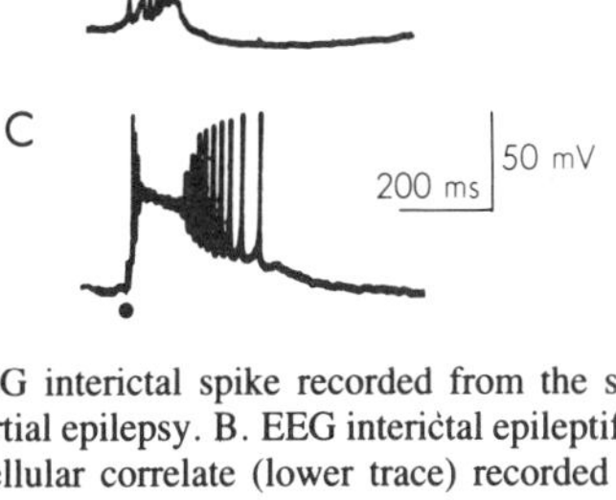

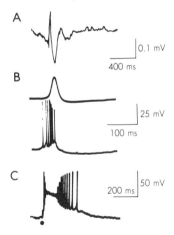

Figure 1. A. EEG interictal spike recorded from the scalp of a patient suffering from partial epilepsy. B. EEG interictal epileptiform spike (upper trace) and intracellular correlate (lower trace) recorded in the cat motor cortex after local application of the epileptogenic agent strychnine; note the burst of action potentials (so-called PDS), which is followed by a long-lasting hyperpolarization. Courtesy of MR Klee, from Fig. 1, p 318, in Purpura DP, Yahr MD, eds (1966): *The Thalamus,* New York: Columbia University Press, reprinted with permission. C. Epileptiform discharge recorded intracellularly in an in vitro slice of human temporal neocortex removed for tumor and bathed with the epileptogenic agent bicuculline.

pattern of EEG waves. In partial seizures the seizure discharge recorded in the EEG starts in a circumscribed region of the head overlying the brain area giving rise to the patient's seizures (the so-called epileptic focus). In generalized seizures the EEG discharges appear simultaneously in all regions of the head. Typically, partial and the "major" form of generalized seizures (i.e., the generalized tonic-clonic convulsion) are followed by postictal depression of cerebral activity during which the EEG first appears flat and then develops abnormal slow waves. There is no such depression after absence attacks, the "minor" form of generalized seizures.

Usually epilepsy does not significantly alter the patient's life expectancy. However, if seizures frequently recur, the condition seriously threatens the prospects for a satisfactory life. If a treatable etiological factor can be identified (e.g., an intracranial tumor), causal treatment is possible and a definitive cure can be achieved. In most instances this is, however, impossible and only symptomatic treatment can be offered by using anticonvulsant medication, which controls seizures in about 60% of epileptic patients. If drug therapy fails and if the seizures are partial, the epileptogenic focus can be surgically excised, provided this causes no neurological deficit. In selected cases the results of this treatment are satisfactory (only very rare or no seizures following surgery) in about two-thirds to three-quarters of patients.

Experimental epilepsy

The mechanisms underlying epileptic discharges have been studied in animal models through a variety of experimental manipulations of brain structure or function. Acute or chronic electrical brain stimulation, topical application of chemical convulsants to cerebral gray matter, systemic application of drugs, physical agents applied to the brain such as freezing—all are capable of inducing electrographic EEG changes resembling both the interictal and ictal EEG patterns seen in human epilepsy, as well as behavioral seizures.

The emphasis in this review is on the electrophysiological mechanisms of epileptic discharge. The important biochemical mechanisms related to these are not covered in detail, mainly because the biochemical studies of epileptic discharge, in contrast to those of its electrophysiological manifestations, have not led to a coherent view of the mechanism of epileptogenesis. Some abnormalities in amino acid metabolism, particularly as it affects the compartmentalization of glutamic acid in intracellular (neurons and glia) and interstitial spaces are found to be associated with many forms of experimental and human epilepsy. It has been suggested that these changes indicate that there is a leakage of glutamic acid from the intracellular to the extracellular compartment where it may exert its well-known excitatory action and thus may be a factor in causing seizures. It is, however, not yet clear whether these changes, rather than being causally related to epilepsy, merely represent a consequence of epileptic neuronal hyperactivity. Other views on biochemical mechanisms involved in epilepsy have stressed a defect in the availability of gamma-aminobutyric acid (GABA), the main inhibitory substance in the brain. GABA levels are, however, often found to be normal in human epileptic brain tissue as well as in that of some animal models of epilepsy.

Experimental models of focal interictal discharge. Most experimental work on focal epileptic discharge has dealt with the mechanism underlying the interictal EEG spike evoked by the acute topical application of chemical convulsants such as penicillin, bicuculline, or strychnine to discrete regions of the brain (mainly the neocortex and hippocampus). Intracellular recordings in an epileptic focus show that at the time of the interictal EEG spike a large amplitude membrane depolarization is associated with a high-frequency burst of action potentials and is usually terminated by a membrane hyperpolarization that may be a classical inhibitory postsynaptic potential (IPSP) or more likely the manifestation of a Ca^{++}-dependent outward K^+ current (Fig. 1B). This phenomenon is known as the paroxysmal depolarization shift (PDS) and is often considered to represent the basic underlying cellular feature of the focal interictal discharge expressed in the EEG as a spike. The latter is assumed to result from the summation of many individual PDSs. A PDS-like phenomenon can also be recorded in neurons maintained in vitro when they are bathed with chemical convulsants (Fig. 1C). Experimental evidence obtained in both in vivo and in vitro preparations suggests that the PDS is dependent upon a network-driven synaptic input, a kind of "giant" excitatory postsynaptic potential (EPSP), though endogenous voltage-dependent currents may also contribute to the depolarizing envelope of the PDS and may play a role in its initiation.

An important factor in the genesis of focal interictal discharge might be a diminished efficiency of Cl^--currents controlled in forebrain neurons by the inhibitory synaptic transmitter GABA. Several chemical convulsants are indeed capable of decreasing the responses of neurons to the iontophoresis of GABA and of abolishing IPSPs. Therefore the increased effectiveness of excitatory inputs underlying the PDS might be caused by disinhibition. A selective loss of neurons containing glutamic acid decarboxylase (the enzyme required for GABA synthesis) has been reported in some experimental brain lesions produced by alumina gel, cobalt, or hypoxia that are known to lead to the development of epileptic seizures. The specificity of these changes with regard to epileptogenesis remains, however, to be demonstrated, and the underlying cellular basis of epileptic discharge in these models is less well understood than in epileptic discharge induced by convulsant drugs.

Experimental models of focal ictal discharge. At times focal interictal discharges induced by topical application of convulsants increase in occurrence and develop into an EEG pattern of focal seizure activity. The cellular correlate leading to ictal activity is a progressive lessening of the post-PDS burst hyperpolarization, suggesting that the occurrence of focal seizures might be caused by further impairment of synaptic inhibition or by a decrease of the post-PDS outward K^+-currents. A role for disinhibition in the genesis of focal ictal discharges is also suggested by the findings that both IPSPs and the responses of neurons to GABA iontophoresis diminish in hippocampal seizures induced by high-frequency electrical stimulation. During experimental focal seizures the membrane potential of the ictally discharging neurons remains depolarized. This may be a factor in permitting excessive Ca^{++}-influx into the discharging cells, which may cause reversible and sometimes irreversible damage to neuronal function and structure.

Focal ictal discharges have also been studied by employing chronic models of experimental epilepsy such as the alumina cream focus or the kindling phenomenon. The latter is an epileptogenic condition induced by repeated electrical stimulation of forebrain structures, such as the amygdala, the hippocampus, or the neocortex. A brief and weak electrical stimulation that, upon its first application, is below the threshold required for producing any detectable electrophysiological or behavioral response will, over time and repeated once daily, begin to elicit focal seizure discharges and ultimately generalized seizures. Finally spontaneous seizures as well as interictal

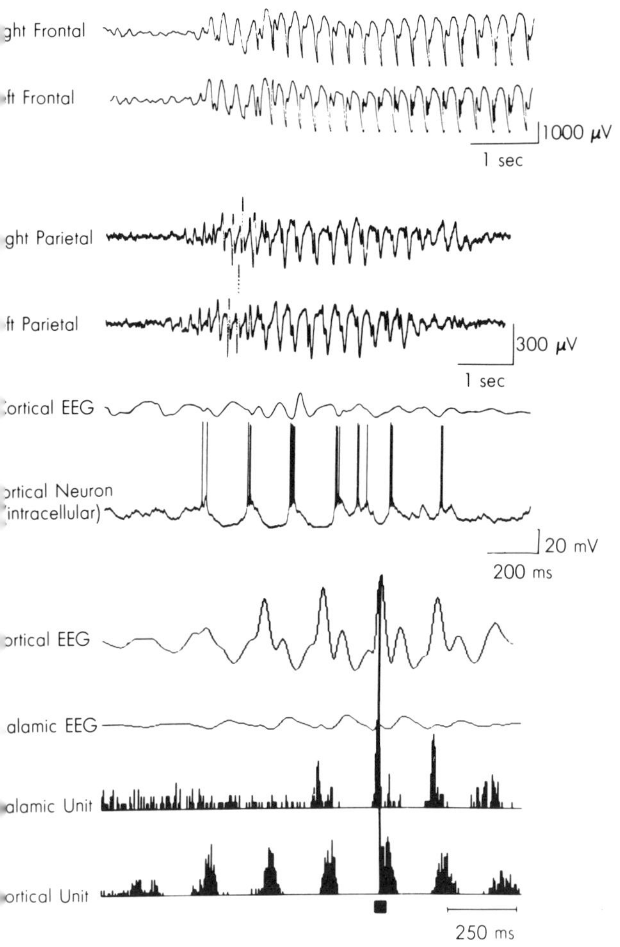

igure 2. A. Generalized spike and wave (SW) discharge from the EEG of a patient during an absence attack. B. Generalized SW discharge induced in a cat by intramuscular penicillin. a. Cortical EEG. b. Intracellular correlate of the SW discharge in a neuron of the motor cortex (lower trace); the EEG showing the rhythmic sequence of SW was recorded from the nearby cortex (upper trace). c. Intracortical EEG averages and time histograms of a cortical and a thalamic cell during generalized SW discharge. The EEG averages were obtained by selecting each of the successive peaks of spikes of SW complexes as time zero (square) for forward and backward averaging of the EEG and for simultaneous computation of the time histograms of action potential discharge (peaks in the histograms are proportional to the number of action potentials occurring at that time; slow wave reversed in polarity because the EEG is recorded intracortically).

EEG spikes occur. Development of kindling is not associated with any detectable morphological changes. There is no evidence for a decrease in the content or in the action of GABA in kindled brain tissue. The electrophysiological basis of kindling remains unclarified. Kindling displays some similarities to long-term potentiation of synaptic transmission in the hippocampus, but there is some evidence that the two phenomena are not directly related.

Focal ictal discharges can spread to distant structures, thus becoming secondarily generalized. Although synaptic mechanisms play an important role in recruiting neuronal elements into participating in epileptiform activity, both at the local circuit level and more distantly, other phenomena such as ectopic action potential generation with antidromic firing along axons, transmission across electrical junctions, and ephaptic effects probably contribute to the high degree of synchronization that characterizes epileptiform discharges.

Experimental models of generalized spike and wave discharges. Generalized bilaterally synchronous spike and wave (SW) discharges in the EEG can be evoked in cats by repetitive low frequency (2–3/sec) stimulation of diencephalic and mesencephalic structures. They are more consistently evoked by the bilateral application of convulsants to the neocortex or by their systemic parenteral administration. For instance, when penicillin is injected intramuscularly in the cat, the resulting generalized SW discharges and the concomitant decrease in behavioral responsiveness closely resemble the electrographic and behavioral features of human absence or so-called petit mal attacks (Fig. 2). These represent the most common minor clinical manifestation of primary generalized epilepsy. Penicillin-induced generalized SW discharges in the cat appear to be generated by the increased responsiveness of neocortical neurons to thalamocortical volleys, which normally evoke spindles. The cellular substrate of the SW discharge is an alternation between periods of increased and decreased neuronal excitation that is sustained within a thalamocorticothalamic loop made more effective by the systemic penicillin (Fig. 2Bc). The period of increased neuronal excitation associated with bursts of action potential discharge corresponds to the spike of the SW complex; the EEG spike may thus represent the sum of synchronous excitatory postsynaptic potentials (EPSPs), while the slow wave component that occurs at the time of the longer period of decreased excitation may result from a summation of the membrane hyperpolarizations recorded intracellularly during the wave of the SW complex (Fig. 2Bb). These, at least in part, represent classical IPSPs. Intracortical inhibitory mechanisms, presumably exerting their action on the neuronal soma, are thus preserved in this model of generalized epilepsy and seem to be an important component in the cellular mechanism underlying SW discharge. The primary feature in its genesis seems to be a state of diffuse cortical hyperexcitability, which can be induced experimentally in cats by the intramuscular injection of penicillin. The powerful activation of intracortical inhibition that occurs under these conditions is probably a secondary effect of the increased discharge of cortical neurons that through recurrent collaterals activates intracortical inhibitory interneurons. The net result is that during SW discharge most of the time is taken up by inhibition. A similar discharge pattern affects thalamic neurons during SW discharge. For generalized SW discharges recorded in the naturally occurring photosensitive generalized epilepsy of the baboon *Papio papio*, a cortical origin has been demonstrated.

Mechanisms of action of antiepileptic drugs

Knowledge of the mechanisms of epileptic discharge is relevant for developing new antiepileptic drugs. Also, knowledge of the mode of action of antiepileptic drugs may make it possible to test some of the current hypotheses concerning the mechanisms of epileptic discharge. Many of the drugs commonly used to treat human epilepsy enhance GABA-mediated inhibition (benzodiazepines, barbiturates). However, some GABA-ergic agents exert a facilitatory effect on experimentally induced generalized epileptiform discharges of the SW type. Recent experimental evidence suggests that some antiepileptic drugs might increase the efficiency of repolarizing K^+-currents (benzodiazepines) and reduce inward depolarizing currents (valproic acid). Both actions may account, at least in part, for their antiepileptic effect.

Further reading

Delgado-Escueta AV, Ferendelli JA, Prince DA, eds (1984): Basic mechanisms of the epilepsies. *Ann Neurol* 16 (suppl)

Jasper HH, Ward AA, Pope A, eds (1969): *Basic Mechanisms of the Epilepsies*. Boston: Little, Brown

Jasper HH, van Gelder N, eds (1983): *Basic Mechanisms of Neuronal Hyperexcitability*. New York: Allan R Liss

Magnus O, Lorentz de Haas AM, eds (1974): *The Epilepsies: Handbook of Clinical Neurology* vol 15, Amsterdam: North Holland Publishing

Penfield W, Jasper H (1954): *Epilepsy and the Functional Anatomy of the Human Brain*. Boston: Little, Brown

Schwartzkroin PA, Wheal H, eds (1984): *Electrophysiology of Epilepsy*. New York: Academic Press

Eye Movement Dysfunctions and Mental Illness

Philip S. Holzman

Disorders of smooth pursuit eye movements have been reported in many patients with functional psychoses. Between 50% and 85% of hospitalized schizophrenic patients and between 20% and 50% of manic-depressive patients show these disorders. In contrast, the prevalence of pursuit dysfunctions in the normal population is about 8%. Patients with nonpsychotic psychiatric conditions such as personality disorders or serious neurotic conditions show a prevalence of smooth pursuit eye movement dysfunctions that is no higher than that found in the normal population. Although such dysfunctions are typically associated with a variety of neurological syndromes, such as Parkinson's disease, multiple sclerosis, and those following hemispheric and brain stem lesions, no obvious central nervous system diseases have been reported in association with the functionally psychotic patients who show eye tracking disorders.

The eye tracking dysfunctions in schizophrenics are not produced by the neuroleptic drugs (phenothiazines, butyrophenones, and thioxanthenes) usually prescribed for psychotic patients, although there is some evidence that lithium carbonate, prescribed for many patients with major affective disorders, produces pursuit dysfunctions. A variety of central nervous system depressants, including alcohol, chloral hydrate, and barbiturates, also disrupt smooth pursuit eye movements.

Inasmuch as schizophrenic patients with abnormal smooth pursuit eye movements are able to execute saccades (or rapid eye movements) with normal latency, accuracy, and velocity, the smooth pursuit dysfunction in these same patients does not reflect poor motivation or inattention, variables that usually interfere with effective and accurate performance by such patients. Although pursuit integrity tends to degrade with increasing age, especially after the age of 50, chronological age does not account for the disorder in psychotic patients, since the greatest majority of patients tested have been in their twenties and age-matched controls have been used.

Identical pursuit dysfunctions have been reported in the first-degree relatives (principally parents and siblings) of schizophrenic patients, but not of manic-depressive patients. These relatives have no history of psychosis or major mental illness. The prevalence of eye tracking dysfunctions within the families of schizophrenic patients suggests that these dysfunctions may be genetically transmitted and represent a biological marker of the vulnerability to schizophrenia. This supposition is strengthened by studies of monozygotic and dizygotic twins who are clinically discordant for schizophrenia; in spite of their clinical discordance, their concordance for eye tracking dysfunctions was over 80% in the monozygotic sets and about 40% in the dizygotic sets.

The specificity of eye tracking dysfunctions for schizophrenia is further bolstered by their frequent appearance in families with a schizophrenic proband (close to 50% of such families have at least one member, who is not the proband, with eye tracking dysfunctions), whereas in the family members of a manic-depressive proband these dysfunctions occur at no greater than normal prevalence. There is, therefore, strong evidence that these eye movement dysfunctions are associated with schizophrenia and tend to occur within families in which there is a member with clinical schizophrenia, associations that suggest a genetic transmission of smooth pursuit tracking disorders. Their presence in a proband, however, cannot be assumed to be pathognomonic of schizophrenia, since they occur in other disorders as well.

A number of schizophrenic patients with unimpaired pursuit movements have parents with eye tracking abnormalities. This suggests the transmission of a latent trait with different manifestations in different persons and in different families. The latent trait, like that for neurofibromatosis, may express itself in the form of manifest schizophrenia, deviant eye tracking, or in both schizophrenia and deviant eye tracking. Identification of those schizophrenic patients and family members in whom the trait is present and those in whom it is absent is crucial to understanding their relationship to schizophrenia.

Smooth pursuit eye movements are measured by asking the subject to follow a target, which is usually a small dot of light that moves in a sinusoidal (pendular) or triangular (constant velocity) wave form. Eye movements are usually recorded by electrooculographic methods or by infrared reflected light techniques. The latter may give more precise representation of the movements of the eyes, although the former is quite acceptable for screening purposes.

There are two types of eye movements involved in this phenomenon. The first is the saccadic, or rapid eye movement, that occurs frequently each day. They are high velocity, high acceleration eye movements, some of them reaching speeds of over 700 degrees per second. Normally about a quarter of a million such movements are made daily. These saccadic movements are active seeking and fixation movements. They direct the eyes to a target by high frequency bursts of nerve impulses to activated agonist and reciprocally inhibited antagonist eye muscles. The second type of eye movement is the smooth pursuit or tracking movement which responds to disparities between target and eye velocities. These pursuit movements are regulated by a dual-mode control system in which the saccadic system places the target on the fovea and the smooth pursuit system keeps the image of the target there.

It appears that the dysfunction in psychotic patients—principally schizophrenics—consists in the replacement of smooth pursuit by saccadic movements (small saccadic jumps that intermittently substitute for smooth tracking) as well as in the intrusion of saccadic eye movements into the pursuit movements. These latter are of various types, some of which are "square wave jerks" or paired saccades that range from smaller than 0.5 degrees to 5 degrees in amplitude; these saccadic intrusions are not necessarily corrective eye movements.

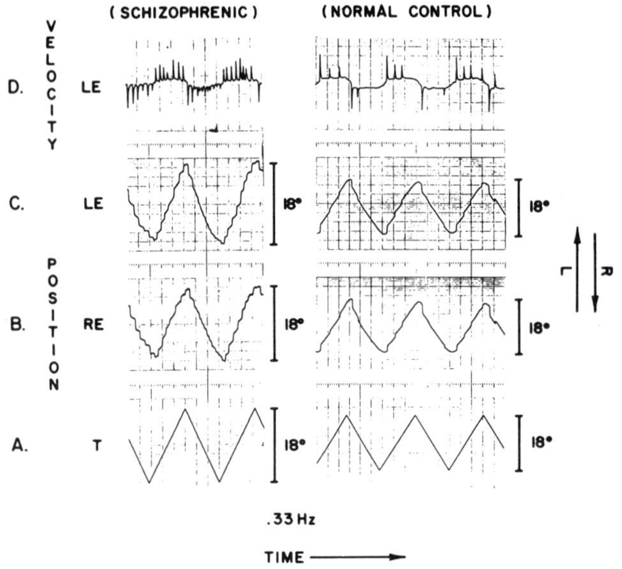

Figure 1. Illustration of a schizophrenic and a normal subject tracking a 0.33 Hz triangular wave. Saccadic intrusions are considerably more frequent in the schizophrenic patient, although they are present in the normal subject. In addition, schizophrenic subjects show a variety of saccadic oscillations.

The former, saccadic tracking, are generally compensatory eye movements that correct for low gain pursuit which causes the eye to fall behind the target; these saccadic substitutions thus aid the eye in recapturing the target. Both types of movements, saccadic tracking and saccadic intrusions are found in some neurological conditions and in some opthalmological diseases, as well as in normal people. Their frequency in psychotic patients, however, is considerably higher than that found in the normal population. Their presence makes it difficult at this stage of knowledge to decide whether the disorder is a failure of the smooth pursuit system or of the modulation or inhibition of the saccadic system, or both. Figure 1 illustrates some of the pursuit patterns found in schizophrenic patients.

Some of the neural mechanisms controlling saccadic eye movements are located principally in the brain stem. Inasmuch as these patients produce saccades with normal trajectories and latencies, those brain stem loci appear to be normal. Since the eye can be made to move smoothly by generating pursuit movements via vestibularly driven eye movements, the extraocular musculature also appears to be normal. Higher centers are likely implicated. Additional evidence for this hypothesis is provided by the demonstration that all psychotic patients who show impaired horizontal tracking when following a moving pendulum show impaired vertical tracking as well; horizontal and vertical tracking controls are located in different parts of the brain stem. Further, these patients show normally smooth eye tracking movements when the eyes are moved by the oculocephalic reflex, full-field (but not partial-field) optokinetic nystagmus, as well as by vestibularly generated smooth pursuit.

The relationship of eye tracking dysfunctions to the symptoms of schizophrenia is not yet clear. Nor is there any obvious relationship between these eye movement disorders in schizophrenic patients and structural brain patterns, although schizophrenic patients with enlarged ventricles on computerized tomographic scans show a high prevalence of these eye movement patterns.

Further reading

Lennerstrand G, Zee D, Keller E, eds. (1982): *Functional Basis of Ocular Motility Disorders*. Wenner-Gren Symposium Series, 37. New York: Pergamon Press

Lipton RB, Levy DL, Holzman PS, Levin S (1983): Eye movement dysfunctions in psychiatric patients: a review. *Schizophrenia Bull* 9:13–32

Stark L (1968): *Neurological Control Systems: Studies in Bioengineering*. New York: Plenum Press

Fetal Alcohol Syndrome

William J. Shoemaker

Although suspected for centuries, the detrimental effects of ethyl alcohol on the unborn fetus were not systematically documented and given the name fetal alcohol syndrome (FAS) until the early 1970s. The characteristics of FAS are central nervous system (CNS) dysfunction, prenatal and postnatal growth retardation, and dysmorphology of the mid-facial region. The appearance of these children resembles those born with Down's syndrome (trisomy 21) because of the short palpebral fissures and prominent epicanthic folds. This facial resemblance may have contributed to the lack of recognition of FAS as a separate condition by the medical profession and the general population; in addition, FAS children are frequently microcephalic and mildly to moderately mentally retarded. The common practice of heavy drinkers to underestimate their alcohol consumption and their reluctance to recognize the contribution of their drinking behavior to their child's condition makes it difficult to assess the quantity of alcohol ingested during pregnancy by mothers of FAS children. Nevertheless, it is clear that FAS children are the offspring of excessive drinkers, many of them alcoholics. What is not clear is whether there is a range of effects on the fetus caused by a corresponding range of alcohol intake. However, danger from moderate use of alcohol (less than 1 ounce absolute alcohol per day) has not been demonstrated and should not be overstated. FAS is now the third leading cause of birth defects (after Down's syndrome and spina bifida) and is, of course, the leading preventable cause.

Much of the research on FAS has focused on the CNS defects. To date there are published neuropathological reports of FAS brains from 16 children. Many of these brains contain glioneuronal submeningeal heterotopias and microscopic abnormalities observed in neurons, but a distinct pattern has not emerged. Many of the FAS children who come to autopsy die of congestive heart failure, a common feature of FAS; moreover, the resulting hypoxia may contribute to the neuropathology observed.

Studies using animal models of FAS have focused on morphological, biochemical, and behavioral indices of prenatal alcohol exposure. Modest deficits in brain serotonin and norepinephrine levels have been described across a number of studies; brain membrane and myelin chemistry, measured by different indices, is also abnormal after prenatal ethanol exposure. Many of the morphological studies have focused on the hippocampus, where aberrant connections and ectopic cells

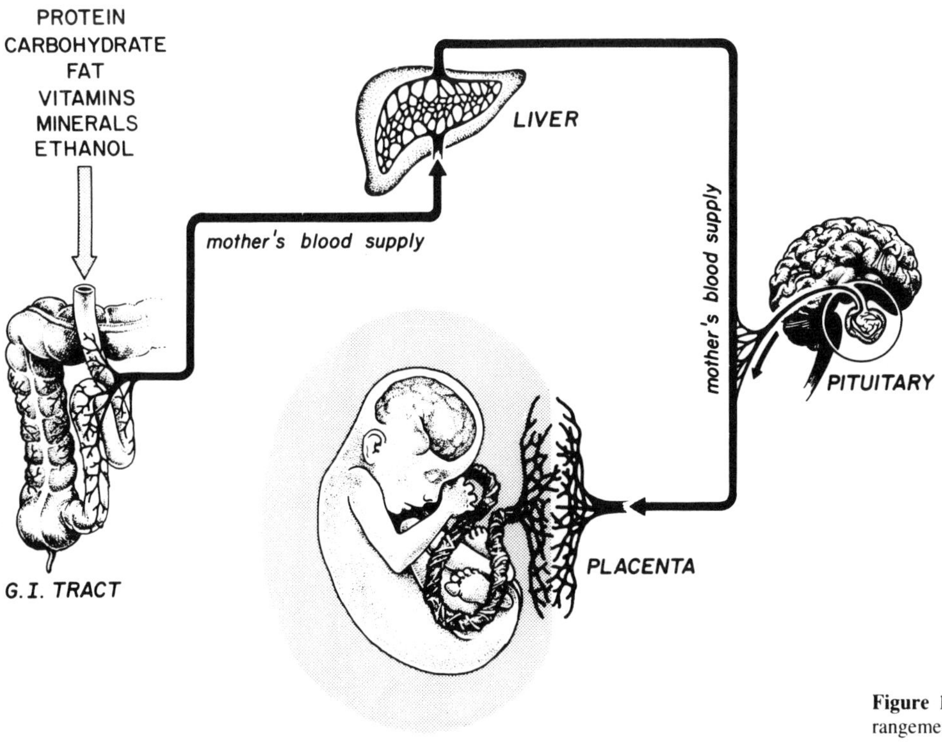

Figure 1. Nutrient, hormonal, and circulatory arrangement in pregnant women.

are described in rodents. These observations support results observed following behavioral testing of adult rodents after prenatal ethanol where memory functions that effect learning behavior have been implicated. One of the more striking effects described in rats exposed to ethanol prenatally is the large increase in the opioid peptide β-endorphin, as well as a decrease in opiate receptors.

Animal studies need to be replicated in other species and expanded in scope before a full picture of the alterations caused by ethanol can be obtained. Nevertheless, certain features can be discerned. Although undernutrition is almost always present in alcoholics, and there is experimental evidence that concomitant undernutrition interacts to make the prenatal effects of alcohol more severe, the biological effects of prenatal alcohol exposure are very different from undernutrition. The complexity and rapidity of cellular events in the developing CNS are such that ethanol's direct and indirect effects would have multiple targets and processes over the course of a 36-week pregnancy. Thus, no single unifying event or process can be discerned in the myriad alterations that have been described. Even the role of ethanol itself as the primary cause of abnormal CNS development is not clear, since ethanol has many targets in the body (e.g., liver, pituitary, placenta) that contribute factors necessary for the growth and development of the fetal nervous system. (See Fig. 1)

Further reading

Ciba Foundation (1984): *Mechanisms of Alcohol Damage in Utero.* Symposium 105. London: Pitman

Rosett HL, Weiner L (1984): *Alcohol and the Fetus: A Clinical Perspective.* New York: Oxford University Press

Shoemaker WJ, Baetge G, Azad R, Sapin V, Bloom FE (1983): Effect of prenatal alcohol exposure on amine and peptide neurotransmitter systems. In: *Drugs and Hormones in Brain Development: No. 9, Monographs in Neural Science*, Schlumpf M, Lichtensteiger W, eds. Basel: Karger

Wisniewski K, Dambska M, Sher JH, Qazi Q (1983): A clinical neuropathological study of the fetal alcohol syndrome. *Neuropediatrics* 14:197–201

Gilles de la Tourette Syndrome

Arnold J. Friedhoff

This disorder was first definitively described in 1885 by Georges Gilles de la Tourette. Symptoms, consisting of multiple tics, inappropriate vocalizations including grunts, coughs, and obscene words, usually begin between 2 and 15 years of age. The first symptoms are often eye blinks, progressing in a fluctuating course to a more generalized disorder including vocalizations. Although the psychological consequences can be severe, this is, at least in part, secondary to the socially intrusive nature of the symptoms. The disorder itself is not believed to produce primary mental problems except for learning disabilities, attentional disturbance, hyperactivity, obsessions, compulsions, and impulsive behavior in some patients. About 40% of sufferers significantly improve in early adulthood, the remainder maintaining symptoms of fluctuating severity throughout life.

Prior to the recognition that this was a specific disorder, victims were often considered to be possessed. In more recent times, because of lack of public and medical awareness of the disorder, patients have often been considered to be deliberately provocative or to be suffering from a mental illness. In either case this has often led to the use of various psychological therapies, which have not been effective. To counter this misconception, victims of this disorder and their families have formed the Tourette Syndrome Association, which has served to raise the level of awareness of the disease and to stimulate research into its origins.

Significant control of symptoms can be obtained in about 75% of cases by the use of the dopamine receptor antagonist haloperidol. The other 25% either do not respond or are intolerant of the side effects of this drug, frequently experiencing sedation, loss of motivation, or extrapyramidal symptoms. Recently another dopamine antagonist, pimozide, has been marketed for treatment and is reported to have less debilitating side effects. The alpha blocker clonidine has also been used experimentally for treatment and is reported to have benefit. Another experimental approach has been through attempted compensatory down-regulation of the dopaminergic system by administering L-dopa, a precursor of dopamine, the beneficial effect occurring when the treatment is stopped. In this approach L-dopa is given to increase the synthesis and release of dopamine at brain synapses. This produces a temporary worsening of symptoms, but ultimately results in a compensatory decrease in the number of dopamine receptors to compensate for the increased release of dopamine. Thus when the L-dopa treatment is stopped net dopaminergic activity is reduced, and symptoms have been found to remit for a time.

Little is known about the etiology or pathophysiology of this syndrome. The disorder follows no known inheritance pattern. It occurs more commonly in boys, and families of patients may have a greater number of members with obsessive compulsive disorder than control populations. Occasionally a so-called high-density Tourette's family has been encountered in which the disorder appears in one or more members across several generations. This would not be expected by chance, because the syndrome is relatively uncommon. Estimates of the exact number of sufferers varies depending on the diagnostic criteria used. Using criteria of early onset, tics, vocalizations, and fluctuating course, as many as 0.05% of the population are afflicted, often being of Mediterranean origin.

Because dopamine receptor blockers are beneficial, it has been proposed that the syndrome involves overactivity of the dopaminergic system; however, at present, there is no direct evidence of such a lesion. It is possible that reducing dopaminergic activity with antagonists such as haloperidol or pimozide may produce a compensatory effect rather than directly impacting on the pathological processes in the disorder.

There is some evidence that Gilles de la Tourette patients suffer from episodic loss of inhibitory control of involuntary movements. Thus motor and vocal impulses would be more likely to be expressed in inappropriate circumstances than suppressed. Obscene words would be selectively expressed because they would most need to be inhibited. It can be readily imagined that a young child finding that "bad" words, coming into his mind, came out of his mouth, would soon become preoccupied with these words, further intensifying the problem. There is, however, no direct evidence for this proposal.

Further reading

Cohen DJ, Shaywitz BA, Young JG, et al (1979): Central biogenic amine metabolism in children with the syndrome of chronic multiple tics of Gilles de la Tourette. *J Am Acad Child Psychiatry* 18:320–341

Friedhoff AJ, Chase TN, eds (1982): *Gilles de la Tourette Syndrome*. New York: Raven Press

Gilles de la Tourette G (1885): Etude sur une affection nerveuse caracterisée par l'incoordination motrice accompagnée de écholalie et de copralalie. *Arch Neurol* 9:19–42, 158–200

Shapiro AK, Shapiro ES, Bruun RD, Sweet RD (1978): *Gilles de la Tourette Syndrome*. New York: Raven Press

Hallucinogenic Drugs

John R. Smythies and C.B. Ireland

There are various types of hallucinogenic drugs that are distinguished by the psychological effects which they induce. The most familiar are the "visionary psychodysleptics" which produce prominent visual and somatic distortions and hallucinations. Complex emotional changes ranging from ecstasy to terror may occur, as well as alterations in thinking. Examples of such drugs are LSD, mescaline, and tryptamine derivatives such as dimethyltryptamine (DMT). The imagery-producing psychodysleptics such as cannabis and various coumarins (e.g., from *Calea zacatechichi*) potentiate visual imagery and produce characteristic distortions in time perception, body-image, and thought. Trance-producing psychodysleptics (such as lysergic acid amide (obtained from *Rivea corymbosa*) ololiquhqui, turbicoryn, and corymbosine), as their names suggest, produce trance-like states. Finally the deliriants (e.g., atropine) produce clouding of consciousness and amnesia, as well as hallucinatory effects.

A second method of classification of the hallucinogens is by chemical formula. There are four main groups: derivatives of (1) phenylethylamine; (2) tryptamine; (3) lysergic acid; and (4) cannabinols. In general, hallucinogens of groups (1) and (2) are methylated derivatives of the brain neurotransmitters, dopamine and serotonin. LSD is a more complex derivative of serotonin, and tetrahydrocannabinol has the same three-dimensional complex molecular shape as LSD (a lipophilic triangle set at right angles on a π cloud). The methylation is on the ring hydroxyls (phenethylamines, e.g., mescaline), side chain nitrogen (e.g., DMT) or both (e.g., o-methylbufotenin, OMB, or 5-methoxytryptamine-DMT).

The mechanism of action of the hallucinogens at the receptor level has been studied intently over the last 30 years. Serotonin (5-HT), dopamine, (DA), histamine, and adrenergic receptors have all been implicated. Aghajanian found that LSD and related indole hallucinogens are potent inhibitors of presynaptic (5-HT) receptors. However, the hallucinogen mescaline does not produce this effect, whereas the nonhallucinogen lisuride does, very strongly. A series of behavioral tests also implicates serotonin receptors in some aspects of hallucinogen action. Other tests involve adrenergic systems, often in complex interrelationships with serotonin systems. Of particular value is the stimulus control technique, in which a rat is trained to react to the injection of a certain hallucinogen by emitting a particular conditioned response. The degree to which injections of other hallucinogens can elicit this conditioned response informs us as to the degree of similarity or difference in their mechanisms of action. Mescaline and certain indolealkylamine hallucinogens (LSD, OMB, 2,5-dimethoxy-4-methylamphetamine) show a good cross-reaction. These reactions are blocked by 5-HT receptor blockers. DA-related compounds (agonists and antagonists) have a much smaller influence on this effect.

However, other tests (e.g., rotational locomotion in 6-hydroxydopamine-lesioned rats) indicate a further action of hallucinogens on dopamine receptors. LSD is a mixed agonist/antagonist at central DA receptors. Many drugs have agonist or antagonist actions at DA receptors, but it has been suggested, as in the case of opiates, that it is the class of mixed agonist/antagonists that produce psychotic reactions. Another hypothesis links hallucinogenic activity to the ability of the compound to have an effect on 5-HT and DA receptors in parallel. In binding studies, it has been shown that LSD binds to 5-HT and DA receptors (to the former in particular in the hippocampus, and to the latter mainly in the caudate). Other evidence suggests that the more potent hallucinogens (such as LSD) decrease impulse flow in 5-HT neurons by 5-HT agonist activity at autoreceptors at the same time as they stimulate DA receptors. Weaker hallucinogens such as psilocyn lack DA receptor agonist activity. There are also marked differences among brain areas. For example, DA-sensitive adenylate cyclase is stimulated by LSD and by mescaline in anterior limbic cortex and the auditory cortex, but by neither in frontal cortex, caudate, or retina.

LSD also binds to alpha-adrenergic and beta-adrenergic receptors, but so does its inactive relative Brom-LSD. Hallucinogens are also competitive antagonists at histamine receptors and the rank order of potency correlates with their clinical rank order. Hallucinogens also have complex interactions with the opiate receptor.

In neurophysiological studies the seminal report was the finding that LSD inhibits the firing of the serotonin-containing cells of the raphe nucleus. Cells in the visual cortex are also very sensitive to LSD, being either stimulated (low doses) or inhibited (higher doses).

Hallucinogens also cause disaggregation of polysomes and increase the acetylation of specific brain histones.

At the molecular level the hallucinogenic potency of a compound has been shown to correlate directly with its ability to act as a charge transfer donor (Domelsmith, 1977) (i.e., by donating electrons from the 4-position on the indole nucleus).

One surprising recent finding has been that the well-known hallucinogens, DMT and OMB, are normal constituents of human bodily fluids such as cerebral spinal fluid (CSF) and rat brain. These compounds appear to be related to stress, since levels are raised in rat brain by stress. CSF, blood, and urine levels are raised in certain psychotic patients, but much larger amounts are found in some cases of liver disease without psychosis, so the significance of this finding is debatable.

The hallucinogens produced by these agents also have features of psychological interest. The simplest manifestations are in terms of Kluver's "form constructs." These start as certain simple geometric patterns such as checkerboards, webs, spirals, etc., that are also characteristic of the stroboscopic patterns. These, however, rapidly become more complex and beautiful, ending up in arabesques, patterns of jewels, etc.

Lastly, fully formed scenes emerge which many observers state have transcendental beauty. The perception of the external world also changes in the direction of increased beauty and significance, as described eloquently by Aldous Huxley in *The Doors of Perception*. There is no particular connection between the content of the hallucinations and the previous experience of the subject. On the other hand, there are close similarities among the hallucinations seen by different people. It has been suggested (Jung, 1952) that these visions originate in the collective unconscious which would explain their quite impersonal nature.

It is important to stress that hallucinogens are indeed dangerous drugs. Psychiatrists are familiar with the acute psychoses that such drugs can induce in many people. Furthermore, these drugs—even in a single experience—can induce painful lifelong symptoms of a psychiatric or a neurotic nature in susceptible people. In this regard marijuana can not only precipitate such acute and chronic psychoses but in addition can often cause a chronic organic brain syndrome with severe cognitive and volitional impairments.

Further reading

Aghajanian GK, Haigler HJ, Bloom FE (1972): Lysergic acid diethylamide and serotonin: direct actions on serotonin-containing neurons in rat brain. *Life Sci* 11:615–622

Domelsmith LN, Munchangen LL, Houk KN (1977): Lysergic acid diethylamide. Photoelectron ionization potentials as indices of behavioral activity. *J Med Chem* 20:1346–1348

Huxley Aldous (1954): *The Doors of Perception*. London: Chatto & Windus

Jung CG (1952): Personal communication

Osmond H, Smythies J (1952): Schizophrenia. A new approach. *J Ment Sci* 98:309

Smythies JR (1959): The stroboscopic patterns. Part I. The dark phase. *B J Psychol* 50:106

Smythies JR, Morin RD, Brown GB (1979): Identification of dimethyltryptamine and O-methylbufotenin in human cerebrospinal fluid by combined GC/MS. *Biol Psychol* 14:549–556

Heroin (Diacetylmorphine)

Conan Kornetsky

" 'H' is for heaven; 'H' is for hell; 'H' is for heroin." So begins the 1964 book, *The Road to H* by Chein et al. This book, published over 20 years ago, is still one of the most comprehensive social-psychological studies of street use of heroin by adolescents. Much of what was written then is still true today. What has changed is the tremendous increase in our knowledge concerning the action of heroin and other narcotic analgesics in the brain.

Heroin (diacetylmorphine) is derived from morphine, the major alkaloid of the opium poppy, *Papaver somniferum*. Heroin is made by exposing morphine to acetic acid, causing a change in the chemical structure of morphine involving acetylation of the phenolic and alcohol OH groups (Fig. 1).

In humans, heroin is metabolized to 6-acetylmorphine and morphine. It is believed that heroin has no activity of its own but that it is a prodrug. That is, all of its pharmacologic action can be accounted for by the action of its metabolites. Binding studies have shown that the metabolites of heroin, but not heroin itself, bind to the opiate receptor. Heroin is a more potent analgesic than morphine. Its onset of action is faster, but it has a shorter duration of action than morphine alone. The reason for these pharmacokinetic differences are believed to be derived from the greater lipid solubility of heroin as compared to morphine and thus its greater ability to pass through the blood-brain barrier.

Heroin was first commercially produced in 1893 and marketed as a cough suppressant. It was clearly more potent than codeine, the opiate derivative usually used for cough suppression. Some early articles suggested that heroin was not addicting, while others raised the possibility that it could cause dependence.

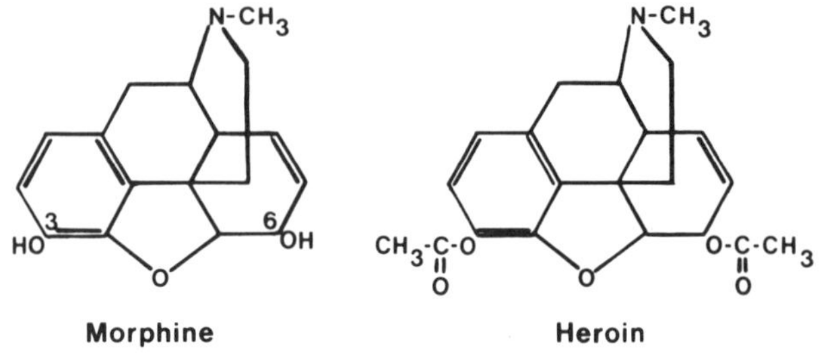

Figure 1. Structural formulas.

At present heroin is not an approved drug for clinical use in the United States, although it is in the United Kingdom. Its main use in the United States is illicit. Although a more potent analgesic than morphine, there is no evidence that it is more effective than morphine in relieving pain. Thus, except for a possible shorter onset of action, the analgesic effects of heroin are believed to be similar to those of morphine. It should be noted that the United States House of Representatives defeated a bill to legalize the use of heroin in cancer patients in August 1984.

Although there is illicit use of morphine, most of the nonmedical use of narcotic drugs is with heroin. The advantage seems to accrue to the illicit manufacturer and seller, for heroin is not only easy to manufacture but because of its greater potency it is easier to transport and smuggle into the United States. There is still belief by some that heroin has some superiority over morphine in causing euphoria. Some recent studies on the relative effects of morphine and heroin on rewarding brain stimulation also suggest that heroin may cause more euphoria than morphine. Although both drugs increase the sensitivity of animals to this rewarding electrical stimulation, the potency of heroin compared to morphine is significantly greater than the reported potency of heroin compared to morphine in relieving pain. If the threshold-lowering effect of the narcotic analgesics on intracranial rewarding brain stimulation is a model for the euphoria caused by these drugs, then it can be assumed that given equal analgesia heroin probably causes more euphoria than does morphine, and this could also contribute to the greater street use of heroin over morphine.

Further reading

Chein I, et al (1964): *The Road to H: Narcotics, Delinquency and Social Policy*. New York: Basic Books

Inturrisi CE, et al (1984): The pharmacokinetics of heroin in patients with chronic pain. *New Engl J Med* 310:1213–1217

Kaplan J (1983): *The Hardest Drug: Heroin and Public Policy*. Chicago: University of Chicago Press

Platt JJ, Labate C (1976): *Heroin Addiction: Theory, Research, and Treatment*. New York: John Wiley and Sons

Huntington's Disease (HD)

James F. Gusella

In 1872, George Huntington, a Long Island physician, drew on his own experience and that of his father and grandfather to provide a graphic description of the inherited neurodegenerative disease that now bears his name. Originally known as Huntington's chorea, the disorder is characterized by progressive involuntary choreiform (dancelike) movements, psychological change, and dementia. It is caused by a dominant gene of as yet unknown function. The onset of symptoms can occur at any time in life but most commonly begins in the fourth to fifth decade. In a small percentage of cases which have been termed juvenile onset HD, the first symptoms occur before age 10. The initial signs of motor disturbance are slight, involving such features as awkwardness of gait, clumsiness, facial twitching, or subtle involuntary movement of the fingers. As the disorder progresses the involuntary movements become more frequent, pronounced, and exaggerated, impairing and eventually eliminating normal day-to-day activity by interfering with the ability to walk, stand, write, speak, and swallow. In some cases, especially those with juvenile onset of symptoms, extreme rigidity is seen rather than chorea. Emotional and behavioral changes often accompany or precede the onset of movement disorder and can similarly have a profound impact on ability to function and on interaction with family members. Impulsive, erratic behavior, impaired memory, poor concentration, and moodiness can all be early signs of HD. In particular, chronic depression occasionally precedes motor symptoms by many years. There is currently no effective treatment for halting or even delaying the progression of HD. Some 15 to 20 years after initial onset of symptoms, the HD victim usually dies of heart disease, pneumonia secondary to aspiration, or choking.

The symptoms of HD are caused by the regional loss of neurons by what has been termed "programmed" cell death. Cell loss is most notable in the striatum but is also seen in the globus pallidus and more diffusely in the cortex. Several different types of neurons have been identified in the striatum based on morphology and neurotransmitters, but the relative rate of loss of these in HD is not known. The HD gene could primarily affect a single cell type whose death leads to the loss of other neuronal elements, or the gene could directly affect several different cell types. Morphological investigations have recently identified specific dendritic abnormalities in spiny neurons of the caudate nucleus in HD although the biochemical basis for this phenomenon is not understood.

The levels of various neurotransmitters have also been investigated in the affected areas of HD brains relative to control brains. Decreases have been recorded in striatal levels of gamma-aminobutyric acid (GABA), acetylcholine, substance P, the enkephalins, and cholecystokinin, likely as a result of, rather than a cause of neuronal loss. Dopamine concentrations have variably been reported as unchanged and increased in HD although dopamine receptor blockers and drugs that deplete dopamine are the most effective palliative therapy for reducing the choreiform movements of HD. Somatostatin concentrations are increased 3- to 5-fold in the caudate, putamen, and globus pallidus of HD brains and may well contribute to the symptoms of the disorder by enhancing the release and actions of dopamine.

Huntington's disease is an autosomal dominant disorder that displays complete penetrance since individuals bearing the defective gene will invariably become affected if they do not die prematurely of other causes. The mean age of onset of symptoms is approximately 38 years, and therefore most HD victims have already had children before the disease appears. On average, 50% of children who have a parent with HD will inherit the disease gene with no bias toward either sex. It has been noted, however, that symptoms tend to appear at an earlier age in children who inherit the HD gene from their father. The disorder differs from some other dominant diseases in its very low mutation rate. In fact, no unequivocal case of new mutation to HD has yet been described. The prevalence of the disorder is approximately 1/15,000 in most populations of Western European origin with much lower frequency in Oriental and pure black populations. At any given time, there are roughly 5 times as many individuals at 50% risk for developing the disease as there are affected, making this a relatively frequent genetic defect considering its severity.

The mode of inheritance of HD, its high penetrance, and the absence of sporadic cases with no family history made this disorder an ideal candidate for the localization of the disease gene by genetic linkage studies. Family studies of this type have recently received new impetus with the application of recombinant DNA technology to generate polymorphic DNA markers that show clear-cut mendelian inheritance by directly monitoring base sequence differences in the genomic DNA. The differences are generally detected as alterations in the pattern of digestion of the DNA by site-specific restriction endonucleases and hence have been termed restriction fragment length polymorphisms (RFLPs). In 1983, the HD locus was mapped to the short arm of chromosome 4 by detection of coinheritance of the disease with such a DNA marker. This represents the first successful application of this new technology to an autosomal disease. It is likely that it will act as a model for the localization of other human disease loci causing such disorders as familial Alzheimer's disease and Von Recklinghausen neurofibromatosis.

The availability of a DNA linkage marker for HD will permit prediction of the inheritance of the HD gene long before symptoms have appeared, and has thereby opened new avenues of research into the genetics and expression of the HD gene. For example, it may now be possible using sensitive noninvasive techniques such as positron emission tomography in conjunction with the DNA linkage marker to determine how long before the appearance of symptoms there is a detectable effect of the gene on metabolism in the central nervous system. One of the major questions concerning the HD gene is whether it is expressed throughout life but only gradually produces the disease phenotype, or whether it is silent for many years and only begins to be expressed later in life. The linkage marker will also permit a better assessment in at-risk individuals of changes of various biochemical parameters that have been reported to be characteristic of HD. Investigations of peripheral tissues in this disease have uncovered a number of possible effects of the HD gene including increased susceptibility of HD fibroblasts to glutamate, sensitivity of HD cells to radiation, alterations in membrane fluidity, and osmotic fragility in HD erythrocytes. To date, many of the potential examples of peripheral expression of the defect have been difficult to replicate and none has yet borne up under intense scrutiny. It should now be possible to assess each of these parameters on a pool of at-risk individuals and only later apply the marker to determine which of the experimental subjects actually possess the HD gene.

The capacity to perform presymptomatic diagnosis in HD will also have a major impact on the potential victims of the disease. It should allow at-risk individuals to plan their lives with a view toward their eventual disability, should they prove to have the HD gene. Similarly, those who have not inherited the defective gene will be free of the constant worry and fear that they might have the disease and have passed it on to their children. Clearly the delivery of such a test presents immense problems since it currently represents a unique situation in genetic counseling in that genetic information with potentially disastrous implications for the individual and his progeny can be obtained many years before there is any sign of the disease. It is likely, however, that the DNA marker linkage technique will soon provide similar capability for a number of other late-onset diseases, including some with significant psychiatric components, and the problems which must be addressed in developing a structure for delivery of a presymptomatic test for HD will have a more general relevance.

The utility of a presymptomatic test for HD would clearly be enhanced if there were a treatment for the disorder, but this is unlikely to be obtained without more information concerning the basis for the programmed neuronal degeneration. Attempts are already under way to use the knowledge of the map position of the HD locus to develop strategies for isolating and characterizing the defective gene and its normal counterpart. The application of recombinant DNA techniques could ultimately lead to an understanding of the nature of the primary defect in this disorder, and lead the way to an effective therapy for halting, or at least delaying, the progression of the degenerative process.

Further reading

Huntington G (1872): *Med Surg Reporter* 26:317–321
Hayden MR (1981): *Huntington's Chorea*. New York: Springer
Gusella JF, Tanzi RE, Anderson MA, et al (1984): *Science* 225:1320–1326
Martin JB (1984): *Neurology* 34:1059–1072

Lithium in Psychiatric Therapy

James W. Jefferson and John H. Greist

While lithium was used in 19th and early 20th century medicine, its ascent to current prominence began in the late 1940s when John Cade, an Australian psychiatrist, noted its antimanic effect. Subsequently, Mogens Schou and colleagues, in Denmark, established that long-term lithium therapy reduced both frequency and severity of episodes in patients with manic-depressive disorder (bipolar affective disorder). These observations were, in part, responsible for a renewed interest in psychiatric diagnosis since distinguishing manic-depressive disorder from the schizophrenias now had important therapeutic implications.

Lithium is currently used in psychiatry to treat a number of illnesses. Its effectiveness is most firmly established in the treatment of acute manic episodes and in the prevention of recurrent episodes of bipolar affective disorder. Indeed, these are the only indications for which labeling approval has been received from the U.S. Food and Drug Administration (FDA).

Both clinical experience and less extensive research support the effectiveness of lithium in some cases of acute unipolar depression, for the prevention of recurrent episodes of unipolar depression, in schizoaffective disorder, in selected cases of schizophrenia (usually in conjunction with an antipsychotic drug), and as an antiaggression agent. The positive response to lithium of some non-manic-depressive conditions suggests that lithium is not specific for manic-depressive disorder and that diagnosis cannot be based solely on lithium response. Some investigators feel lithium is useful for treating alcoholism either indirectly by stabilizing mood or directly by reducing alcohol consumption, but this remains controversial. While lithium has been investigated in a variety of other psychiatric disorders, its efficacy has not been established.

The impact of lithium therapy on the course of bipolar affective disorder is striking. Untreated, the illness is characterized by recurrent episodes of mania and depression, often lasting many months and often associated with marked disability and sometimes death (suicide, aggravation of medical conditions, reckless behavior). The Medical Practice Information Demonstration Project estimated that without treatment the average woman with the onset of the disorder at age 25 would lose 9.2 years of life, 14.2 years of major life activity, and 11.9 health status years. With optimal treatment these losses would be reduced to 2.7, 4.0, and 3.4 years. The economic impact of lithium treatment for manic-depressive disorder in the United States has been conservatively estimated to result in a savings of over $400 million per year (reduced medical costs and increased productivity).

Pharmacology

Lithium is the simplest solid element (atomic #3, atomic weight 6.94) and a member of Group IA of the periodic table (together with sodium, potassium, rubidium, cesium, and francium). It also has chemical similarities to Group IIA elements such as calcium and magnesium. In the body, it is neither metabolized nor protein bound, is rapidly and fully absorbed from the gastrointestinal tract, and is excreted almost entirely by the kidneys.

Mechanism of action

Despite its simple chemical structure, lithium exerts profound effects throughout the body at levels ranging from cellular to behavioral. Extensive investigations have attempted to explain its remarkable clinical effects, and while much is known about how the drug alters hormonal, neuronal, and metabolic systems, its exact mechanism of action is unknown. It does seem clear that bipolar affective disorder is not caused by lithium deficiency and that lithium does not exert its effect by correcting such a deficiency. While a lithium deficiency state has not been experimentally produced in man, the naturally occurring levels in blood (about 9 ng/ml) are about 600–800 times lower than those required for therapeutic effectiveness, suggesting that lithium works as a drug rather than as a mineral replacement.

Given its similarities to critical body cations such as sodium, potassium, calcium, and magnesium, lithium may work through ion substitution. Studies have shown that lithium exerts effects on neurotransmitters such as norepinephrine, serotonin, and acetylcholine in ways that are consistent with current theories of affective disorders. Lithium also inhibits the activity of the enzyme adenylate cyclase, thus interfering with the action of vasopressin on the kidney and leading to polyuria in a substantial number of patients. Inhibition of beta-adrenoceptor-stimulated adenylate cyclase or prostaglandin-E_1-stimulated adenylate cyclase has also been postulated as a mechanism of action.

Lithium inhibits replication of DNA viruses such as types 1 and 2 herpes simplex, and preliminary evidence suggests a use in treating herpes genitalis and labialis. Of greater and even more speculative interest is the similarity between the episodic clinical course of both herpes simplex and manic-depressive disorder and the possibility that lithium works for the latter through inhibiting an as yet unidentified virus.

Monitoring therapy

The chemical simplicity of lithium makes it easily measured in the blood, and unlike other psychiatric drugs, monitoring of blood levels has become an integral part of lithium therapy. Both flame photometry and atomic absorption spectrophotometry are key techniques readily available in most clinical laboratories and well suited for reliable and accurate measurement of serum lithium levels. Of research interest are more sensitive methods such as flameless atomic absorption spectroscopy, inductively coupled argon emission spectrometry, and mass spectrometry for measuring lithium in biological fluids. When used therapeutically, lithium is present in serum in relatively high concentrations so that the conventional techniques are more than adequate. While lithium has been measured in virtually every body fluid, there is little likelihood that blood will be replaced as the standard of therapeutic monitoring.

Several lithium preparations are available in the United States. These include standard and slow-release lithium carbonate in the form of capsules or tablets and the liquid, lithium citrate. A proper dosage of lithium is established by both evaluating the patient's clinical status and adjusting the serum level to a range of 0.8–1.2 mEq/L for mania and 0.6–1.0 mEq/L for maintenance therapy. For those values to be meaningful, lithium must be given in divided doses and blood samples must be drawn in the morning as close as possible to 12 hours after the last dose. Even then, these ranges are only guidelines, and patients may require higher levels for benefit or tolerate lower levels without relapse. There is some evidence that a substantial number of patients can be successfully maintained at serum levels of 0.4–0.6 mEq/L. Unfortunately, there is no reliable way to predict in advance who these patients will be. The risk of lithium toxicity increases as the serum level rises, and levels above 1.5–2.0 mEq/L are likely to be dangerous.

Adverse reactions

Side effects of lithium range from benign and tolerable to life-threatening and affect every organ system. Both central and peripheral neurological function can be altered by lithium. The electroencephalogram (EEG) may show increased amplitude and generalized slowing with reduced alpha and increased theta and delta activity, and occasionally focal changes occur. Somatosensory and auditory evoked potentials may also be increased. In therapeutic amounts, patients may experience tremor, fatigue, and slowed information processing. Lithium intoxication is predominantly a neurotoxicity which may initially present as lethargy, dysarthria, and ataxia but which can progress to extreme neuromuscular irritability, seizures, coma, irreversible neurological damage, and death. Although the therapeutic index of lithium is low, proper clinical and laboratory monitoring make severe intoxication uncommon. The drug is most rapidly and effectively removed from the body by hemodialysis which is the treatment of choice for severe intoxication.

Lithium has complex effects on thyroid function, the most consistent of which is to inhibit release of T3 and T4 from the gland. Other effects include inhibition of thyroglobulin biosynthesis, inhibition of thyrotropin (TSH)-sensitive adenylate cyclase, and inhibition of iodine uptake and organification. The net clinical result is the development of goiter and hypothyroidism in a small percentage of patients. The periodic assessment of thyroid function is, therefore, an integral part of monitoring therapy. Blood levels of T3, T4, and TSH are used routinely while the more sensitive but costly thyrotropin-releasing factor (TRF) infusion test remains of research interest.

Calcium metabolism and parathyroid function are also affected by lithium, possibly by altering the parathyroid gland's "set point" so that higher calcium levels are required to inhibit parathyroid hormone secretion. The end result is an increase in serum calcium and parathyroid hormone levels.

While lithium has been shown to have a variety of effects on carbohydrate metabolism at both a hormonal and cellular level, its relationship to diabetes mellitus remains unclear. The weight gain which often occurs during lithium maintenance therapy may be, in part, due to altered carbohydrate metabolism.

Lithium stimulates granulocyte production, possibly through enhanced production of colony-stimulating factor, with a resultant increase in white blood count. This "side effect" has been utilized to treat granulocytopenic conditions such as Felty's syndrome, idiopathic neutropenia, and chemotherapy-induced neutropenia. An increase in circulating platelets has also been noted.

The effects of lithium on the kidney are well established and clinically important. A nonspecific chronic interstitial nephritis occurs more commonly than in control patients and a reversible lithium-specific lesion has been described. These structural changes are most notable in the distal convoluted tubules and collecting ducts. Changes in renal function induced by lithium include a reduction in renal concentrating ability which appears to be a reflection of tubular interstitial nephropathy, and an increase in urine volume. The latter is due primarily to reduced renal response to ADH (vasopressin) at the level of renal adenylate cyclase but also may involve impaired utilization of cyclic AMP, impaired proximal tubular fluid reabsorption, and a central effect on synthesis, storage, and release of ADH.

Lithium also impairs renal hydrogen ion secretion, resulting in a partial distal tubular renal acidosis which does not appear to be of clinical importance. Finally, the effect of lithium on glomerular filtration appears to be modest, and there is little evidence to suggest that prolonged therapeutic exposure to lithium results in renal failure.

Lithium information

In 1975, the Lithium Information Center was established at the University of Wisconsin so that medical and allied professionals throughout the world could have convenient access to a comprehensive collection of bibliographic references to the medical and biological lithium literature. As of early 1986, over 14,000 articles have been entered in a computer-based storage and retrieval system which can be rapidly searched by key word, title, author, and journal in a way that is both complete and specific. The Lithium Information Center is a prototype for specialized information centers that will eventually provide a networking of computer-accessible knowledge systems.

Further reading

Jefferson JW, Greist JH, Ackerman DL (1983): *Lithium Encyclopedia for Clinical Practice*. Washington: American Psychiatric Press

Johnson FN, ed (1980): *Handbook of Lithium Therapy*. Lancaster, England: MTP Press

Lazarus JH (1982): Endocrine and metabolic effects of lithium. *Adv Drug React Ac Pois Rev* 1:181–200

Knapp S (1983): Lithium. In: *Psychopharmacology 1, Part 1: Preclinical Psychopharmacology*, Grahme-Smith DG, Cowen PJ, eds. Amsterdam: Excerpta Medica

Marijuana

Robert C. Petersen

Marijuana, a mixture of the leaves, flowering tops, and other parts of the plant Cannabis sativa, is a chemically complex substance. It contains over 400 chemical constituents including simple acids, alcohols, more complex hydrocarbons, terpenes, and some 61 compounds unique to the cannabis plant called cannabinoids. The cannabinoid Δ-9-tetrahydrocannabinol (THC) is the principal psychoactive ingredient, although other constituents may modify its action or have psychoactive effects of their own. This compound is present in varying amounts. High-grade seedless varieties of cannabis called sinsemilla may have percentages (by weight) of THC as high as 11%. Street marijuana is believed to have increased in average THC content—from less than 1% in the 1960s and 1970s to as high as 5% in the mid-1980s. Variability in marijuana's potency and a failure to adequately specify the material used has often made the results of earlier marijuana research (before 1960) hard to interpret. Cannabis cigarettes supplied by the National Institute on Drug Abuse for research use typically contain 1–2%.

Cannabis has been the source of fiber used in making rope and coarse woven cloth for thousands of years. Knowledge of its putative medicinal and mind-altering properties is also ancient. In modern times, concern about marijuana's possible abuse and potential health hazards dates back to at least the 1890s when an Indian Hemp Drugs Commission was appointed to study its effects in India where use was—and still is—traditional.

While cannabis was widely cultivated for its fiber content in North America beginning in Colonial times, there is little evidence of its widespread use as a drug prior to the turn of the century. Up until the 1960s use was largely restricted to a small number of adults. A marked increase in use has occurred since, mostly among adolescents and young adults. By the 1980s three out of five American youths between 18 and 25 had tried the drug with a third continuing to use it with some regularity.

Acute effects

Acute effects of marijuana intoxication (or "high") range from subjectively experienced increase of esthetic sensitivity and heightened sexual sensation to mild tachycardia and other measurable changes in performance. A sense of relaxation and tranquillity is commonly reported, but heightened levels of anxiety and paranoid feelings are also common. There is little question that marijuana at common dosage levels interferes with many aspects of psychomotor and cognitive function in ways that adversely affect learning and skilled performance while high. Dozens of experimental studies confirm that such cognitive processes as arithmetical reasoning, verbal and nonverbal problem solving, and short-term memory are adversely affected. Marijuana's effects on immediate memory and an increased distractability while high may be common elements. Complex performance involved in operating a car or an aircraft has also been found to be significantly impaired by marijuana intoxication. As with alcohol, there is good reason to believe that the deficiencies measured in the laboratory, on the test course, or in the driving simulator prove the hazards of use to highway safety.

Chronic effects

While many of marijuana's acute effects have been explored, possible chronic effects are far more difficult to determine. American experience with cannabis has been brief and largely confined to the healthiest segment of the population. Research on traditional users in countries like India or Jamaica may have only marginal relevance to contemporary American patterns. Traditional users are usually adult males whose mode and circumstances of use differ significantly from the newer use patterns. For example, since cannabis is often mixed with tobacco when smoked by traditional users, deep inhalation may be less common, making the apparently large quantities used deceptive. Extensive use by children and adolescents during critical stages of physical and psychological development may have quite different, and more serious, implications than those for adults, whether or not the drug is used traditionally.

Respiratory effects. Since marijuana is usually smoked and the smoke deeply inhaled, effects on respiratory function would not be surprising. Although not usually consumed in the same quantities as tobacco cigarettes, marijuana "joints" are consumed almost entirely, the smoke typically unfiltered and each puff retained in the lungs for several seconds. Research on healthy young adult male volunteers in a chronic marijuana smoking study has found mild, but measurable evidence of airway obstruction after one and a half to two months of smoking five cannabis cigarettes daily. Studies exposing animals to marijuana smoke which produced similar blood cannabinoid levels to those in human daily users have found degenerative lung changes following periods of use corresponding to an eighth to a half the animal's life span. Clinicians who have studied heavy American users of hashish overseas have found cellular changes in bronchial tissue similar to those in heavy cigarette smokers, but at younger ages. Studies of isolated human lung tissue cultures have found precancerous changes in lung tissue exposed to marijuana smoke. When marijuana smoke residuals have been skin tested in animals, tumors similar to those produced by tobacco tar have resulted. The brevity and generally low level of American exposure to marijuana make it unlikely that marijuana-related cancer, emphysema, or other serious pulmonary consequences would be detectible by large-scale epidemiological studies at this time.

Cardiovascular effects. Although temporary increases in heart rate and increased blood concentrations of carboxyhemoglobin following marijuana use have been viewed as generally benign in young users, these effects may be more serious in those with already impaired cardiovascular function. The effects of smoking a marijuana cigarette are greater than those of a tobacco cigarette in producing exercise-induced angina in patients with impaired cardiovascular circulation.

Reproductive effects. Reductions in testosterone levels and reduced sperm counts in human males have been reported by some investigators. Although the values were still within normal limits, the reductions may have significance for the marginally functioning. Effects on human females are still less certain. A possible reduction in fertility has been reported in one study. A higher incidence of the fetal alcohol syndrome in infants born of mothers who used marijuana during pregnancy has also been reported, but the difficulties of separating the effects of marijuana from those of concurrent alcohol and tobacco use make the role of cannabis uncertain.

Brain damage and persistent personality alteration. The question whether chronic heavy marijuana use causes brain damage is not easily answered. Although one poorly controlled human study and a preliminary animal study have suggested the possibility, two more carefully controlled human studies using computerized tomographic techniques have found no evidence of this in chronic users. However, long-term daily administration to laboratory animals has been reported to cause possibly permanent effects on learning of various types. A loss of conventional motivation, the ''amotivational syndrome'' has been clinically described in some heavy users. It is, however, often difficult to distinguish the role of marijuana from that of other drug use and preexisting motivational or psychological problems. A third of a national sample of high school seniors using marijuana daily reported loss of interest in non-drug-related activities and impaired school or job performance that they attributed to marijuana use.

A variety of other problems have been linked with marijuana use, including possible chromosome abnormalities, alterations in cell metabolism, and impairment of the immune response. In these areas research to date has resulted in conflicting findings and must be regarded as inconclusive.

Extensive research on possible adverse consequences of marijuana has also led to renewed interest in its possible therapeutic applications. Thus far the most promising of these has been as an antiemetic in controlling the nausea and vomiting that accompany cancer chemotherapy. Marijuana or its principal psychoactive ingredient have also shown some usefulness in glaucoma treatment. While the drug reduces intraocular pressure by a different physiological mechanism from that of other drugs used in glaucoma treatment, its side effects on memory and other psychological functioning are more often disturbing to elderly patients. Other uses that have been suggested or considered have either not been adequately assessed or have not lived up to preliminary expectations.

Despite the marked increase in research on marijuana, many important questions regarding its effects remain unanswered. Most crucial of these are the implications of frequent use for the developing child or adolescent and the chronic effects of long-term heavy use. While there is little hard experimental data in these areas, clinical observations suggest that there is a serious basis for concern.

Further reading

Hollister LE (1984): Health aspects of cannabis use. In: *The Cannabinoids: Chemical, Pharmacologic and Therapeutic Aspects,* Agurell S, Dewey W, Willette R, eds. New York: Academic Press

Institute of Medicine (1982): *Marijuana and Health.* Report of a Study by a Committee of the Institute of Medicine, Division of Health Sciences Policy. Washington DC: National Academy Press

Petersen RC, ed (1980): *Marijuana Research Findings: 1980.* National Institute on Drug Abuse Monograph Series. Washington DC: US Government Printing Office

Mental Illness, Genetics of

Steven Matthysse

Only adoption studies can provide solid evidence for the existence of heritable factors. Twin studies, showing that the concordance rate in monozygotic twins exceeds that in dizygotic twins, are weakened because monozygotic twins are likely to be treated more similarly by significant persons in their environment, and may also tend to identify psychologically with each other. Studies of family process, showing that families with disturbed communication patterns are more likely to have a schizophrenic child, are not compelling evidence for nongenetic factors, because the disturbed behavior of the parents could be a consequence of the behavior of the child, or a parallel effect of the same genes that caused schizophrenia in the offspring. There are, however, adoption studies in both schizophrenia and affective disorder that point toward the existence of heritable factors.

The most comprehensive adoption study of schizophrenia is that of Kety, Wender, Rosenthal, Schulsinger, and associates, beginning in 1963 in Denmark. Out of a total national sample of adoptees, 17 chronic schizophrenics were selected, and normal controls were matched for age, sex, socio-economic status of the adopting family, and time spent with the biological parent (which in all cases was brief). The biological relatives of the index cases included 5 chronic schizophrenics (7.1%), compared to none in the controls; and 12 "latent schizophrenics" (17.1%), compared to 6 in the controls (6.1%). There was also an increased prevalence in the half-siblings where the shared parent was the father, so that the proband and the affected relative had no common maternal influence, even during gestation.

Various objections have been raised to these studies: (1) There was no estimate of interjudge reliability in diagnosis. (2) The rate was higher in second-degree than in first-degree relatives, whereas first-degree relatives ought to be at greater risk, in a genetic theory. (3) The "latent schizophrenics" were not really schizophrenic at all. (4) Certain diagnoses (inadequate and schizoid personality) were excluded from the "latent schizophrenia" category after the data were collected, inflating the significance levels. (5) The five chronic schizophrenics were in just two families. (6) For statistical purposes, families should be counted, not individuals, since, except for the index case, family members were not separated from each other.

None of these objections is compelling. Poor interjudge reliability would militate against finding significance, not create it if the null hypothesis were true. The higher rate of illness in second- than in first-degree relatives can be accounted for by the nature of the sample. Since the cause of the adoption, in most cases, was illegitimacy, there were almost no full siblings, and nearly all the first-degree relatives were parents. Parents would be expected to have a lowered rate of schizophrenia, because schizophrenic patients are less likely to have children.

An increased risk for "latent schizophrenia" in families of schizophrenic probands is evidence for genetic transmission, even though the illness is milder. Indeed, an increased probability of any trait in relatives of probands adopted at birth (relative to matched controls) is evidence for the operation of transmissible factors in the probands' illness. The investigators maintain that the exclusion of inadequate and schizoid personalities from the "latent schizophrenia" category was part of the original design. Counting families may be a safer way to estimate significance levels than counting individuals, but the investigators analyzed their data both ways, and the results remained significant. The remaining objection, that the five chronic schizophrenics were found in only two families, is a valid observation, but the investigators did not claim that all forms of schizophrenia are equally heritable; they view the illness as probably heterogeneous. (Kety, 1983, cited below, summarizes the Danish adoption study data with response to critics.)

There also are adoption studies in affective disorder. Although on a smaller scale than the Danish schizophrenia study, the results support the existence of heritable factors, especially in severe depression and manic-depressive illness. There is evidence also that suicide has a strong genetic component. Out of the 57 biological families of adoptee probands who had committed suicide, suicide occurred in 11 families (12 relatives), whereas in the control group there were only two suicides among the relatives, both in the same family.

The most rapidly growing aspect of human genetics is gene mapping through linkage to restriction fragment length polymorphisms (RFLP). The discovery of genetic linkage—co-segregation with a Mendelian marker trait within families—provides unequivocal evidence for genetic transmission. It also demonstrates that a major locus is involved; it potentially offers the opportunity for prenatal detection; and it can ultimately lead to identification of the gene, by mapping to a well-understood region of a chromosome.

The technique rests upon the existence of certain enzymes that cut DNA at points where they encounter precisely specified sequences. Variations in the DNA sequence from individual to individual will cause the cuts to occur at different places, resulting in fragments of different length. These variations can be used for linkage analysis in just the same way that traditional blood group markers of HLA-polymorphisms are used. They have the advantage that sequence variation in the regions of DNA that are not expressed in the final protein ("introns") can also provide linkage markers. The noncoding regions may be more variable than the coding regions, because of diminished pressure from natural selection.

The recent success of RFLP methods in mapping the gene for Huntington's disease to the short arm of chromosome 4 by J. Gusella and co-workers has drawn the attention of investigators in psychiatric research. To detect polymorphisms, this group used a recombinant phage probe, G8, derived from a human gene library and mapped to chromosome 4. The restriction fragment data suggested the existence of two "sites"—

small regions of DNA whose exact sequences determine susceptibility to cutting by the restricting enzyme—which are polymorphic (variable) in the population. In an American and in a very large Venezuelan pedigree, Huntington's disease cosegregated with the G8-detectable restriction sites.

The major difficulty in applying this powerful technique to psychiatric disorders is that they are certainly not inherited as classical Mendelian traits, which is the context in which linkage analysis developed. The incidence of schizophrenia in the first-degree relatives of schizophrenics—3.2%, according to Tsuang—is far too low for classical Mendelian inheritance.

Whenever a disease has non-Mendelian transmission characteristics, the possibility has to be considered that the disease itself is not the most appropriate trait for genetic investigation. Partial penetrance is not an explanation, but a sign that the list of causal factors is incomplete. To speak of vulnerability as inherited only obscures the task of elucidating the actual traits that are inherited and of finding out how they create a predisposition to disease.

Two recent studies of schizophrenia have adopted this point of view. L.A. De Amicis and R.L. Cromwell measured reaction time crossover in schizophrenic patients (a paradoxical lengthening of reaction time when warning cues are given at a constant interval before the test stimulus, frequently found in schizophrenics, but rarely in normals). Their first-degree relatives had a tendency to show crossover, even though they had no psychotic symptoms. P.S. Holzman examined smooth-pursuit visual tracking, which is disturbed in about 50% of schizophrenic patients, but only infrequently in normals. Schizophrenic patients who had no disturbance in visual tracking were also included in the sample, and their relatives tended to have bad tracking even if they were nonpsychotic.

It may turn out that no single trait, detectable at the present state of technology, is the sought-for Mendelian risk factor, but that an unobservable or latent trait with Mendelian characteristics underlies both the disease in question and the other traits that seem related to it. A neurological disease process might produce unpredictable combinations of symptoms, depending on where it happened to strike in the nervous system. The vulnerability of different target areas might be influenced by multigenic factors distinct from the disease process itself. For example, smooth-pursuit visual tracking disturbance and schizophrenia might both be manifestations of a yet unknown latent trait, and this trait might be determined by a major gene with higher penetrance than either smooth pursuit or schizophrenia. There are methods for using latent traits in genetic analysis, even if their concrete biological interpretation is still unknown.

The obstacles to discovering linkage for non-Mendelian traits are much greater than for traits that obey the classical laws of genetic segregation. Reduced penetrance (or multigenic effects, which have the same consequences for linkage analysis) decreases the probability of success.

Heterogeneity is also an obstacle to linkage analysis. If very large families are studied, each one is a universe unto itself, and heterogeneity among families does not prevent the discovery of linkage in a subset. If only small families are available, however, there is a high risk that the sum of the lod scores (logarithm to the base 10 of the odds in favor of linkage) across families will not reach significance, even if linkage is present in some of them. The more that can be done to subdivide the trait or illness into etiologically homogeneous subtypes, the greater the likelihood linkage analysis will succeed.

One centimorgan (cM, the chromosomal map distance corresponding to 1% probability of recombination at meiosis) is the approximate lower limit of resolution that can be achieved by linkage techniques in human pedigrees. This level of resolution remains coarse from a genetic point of view. One centimorgan corresponds to approximately 1 million base pairs, or 1,000 genes. If there is some knowledge of the anatomical pathways and physiological mechanisms underlying the trait for which a genetic locus is sought, functional considerations can be used to narrow the search for the gene. Either cDNA or peptides coded by each gene can be used to construct histochemical stains, and if these stains show that the gene product is localized to brain regions known to be relevant to the trait being mapped, the likelihood that the right gene has been found will be enhanced.

Hybridization and peptide immunohistochemistry are both powerful techniques. In a hybridization study by Hudson and co-workers, an 800-base pair DNA fragment corresponding to the structural gene for rat growth hormone was cloned and used as a template for synthesis of labeled cDNA. Hybridization to the cDNA revealed specific labeling in the anterior pituitary. Sutcliffe and co-workers selected cDNA clones hybridizing only to brain mRNA. After sequencing them, they translated the open reading frames into amino acid sequences (according to the genetic code), and chemically synthesized short peptides within those sequences. Antibodies were raised against the synthetic peptides, and these were made into histochemical stains by the indirect immunoperoxidase method. One such stain, derived from a cDNA clone whose function was originally unknown, reacted with a new fiber system, widely distributed in the brain. Viral and plasmid vectors can also be used to synthesize peptides (or antigenic sub-sequences of peptides) from DNA clones.

The success of these molecular techniques depends, in an essential way, on the progress of neuroscience in localizing the pathways, cell types, organelles, and macromolecules involved in human traits. An increasing partnership between genetics and neurobiology is to be expected, as scientific concerns focus more and more on molecular mechanisms in mental disease.

Further reading

Botstein D, White RL, Skolnick M, Davis RW (1980): Construction of a genetic linkage map in man using restriction fragment length polymorphisms. *Am J Hum Genet* 32:314–331

Gottesman II, Shields J (1982): *Schizophrenia: The Epigenetic Puzzle.* New York: Cambridge University Press

Kety SS (1983): Mental illness in the biological and adoptive relatives of schizophrenic adoptees: Findings relevant to genetic and environmental factors in etiology. *Am J Psychiatry* 140:720–727 (Contains a summary of the Danish adoption study data, with responses to critics.)

Kidd KK, Matthysse S (1978): Research designs for the study of gene-environment interactions in psychiatric disorders: Report of a Foundations' Fund for Research in Psychiatry panel. *Arch Gen Psychiatry* 35:925–932

Mendelwicz J (1981): Adoption study in affective illness. In: *Biological Psychiatry,* Perris C, Struwe G, Jansson B, eds. New York: Elsevier

Schulsinger F, Kety SS, Rosenthal D, Wender PH (1979): A family study of suicide. In: *Origin, Prevention, and Treatment of Affective Disorders,* Schou M, Stromgren, E, eds. New York: Academic Press

Sutcliffe JG, Milner RJ, Shinnick TM, Bloom, FE (1983): Identifying the protein products of brain: Specific genes with antibodies to chemically synthesized peptides. *Cell* 33:671–682

Tsuang MT, Winokur G, Crowe RR (1980): Morbidity risks of schizophrenia and affective disorders among first degree relatives of patients with schizophrenia, mania, depression and surgical conditions. *Br J Psychiat* 137:497–504

Mental Illness, Nutrition and

John W. Crayton

Long-standing public interest in diet and nutrition as a means to better mental health has been matched in the last 10 years by a rapidly expanding scientific interest in the relationship between food and behavior. There are, however, fundamental research problems in this area: (1) accurately assessing an individual's nutritional status is difficult; (2) nutritional requirements and response to nutrients vary widely from person to person; (3) the frequently subtle effects of nutrients, or lack thereof, on behavior defy reliable and reproducible measurement with currently available behavioral assessment instruments; (4) brain mechanisms underlying food-behavior phenomena are largely unknown.

Developmental aspects

Generalized maternal malnutrition results in a decrease in both the number and size of individual nerve cells in experimental animals. The behavioral effects of this nerve loss may be broad and difficult to measure, but there is evidence from human studies that maternal malnutrition may have significant effects on offspring behavior. The developing fetus is vulnerable to several specific maternal nutritional deficiencies suffered during the periconceptual phase (preconception to implantation). Iodine-deficient mothers may produce offspring with cretinism. Insulin-dependent diabetic mothers whose diabetes is not well controlled may produce offspring with a variety of congenital malformations or mental retardation. Mothers with untreated phenylketonuria may produce mentally retarded offspring. Preliminary evidence suggests that mothers who have previously given birth to children with neural tube defects have a reduced incidence of this defect in subsequent offspring if they take vitamin B_{12} supplements prior to conception.

The period of fetal development from implantation to the 12th week postconception is a vulnerable period in particular for the offspring of protein-calorie malnourished mothers. Studies of victims of the World War II Dutch famine suggest that children born of mothers suffering protein-calorie malnutrition during this phase had a high incidence of prematurity, stillbirths, and neonatal deaths. Long-term follow-up of surviving children revealed a high incidence of neural tube defects, cerebral palsy, and obesity.

Malnourishment after the first trimester is less critical for the development of optimal brain development, and the effects are more subtle and more subject to reversal by subsequent nutritional supplementation. Studies of Korean orphans malnourished for various periods of time postnatally indicate that those children renourished before the age of 3 fared better than children renourished later. These studies also suggest that the psychosocial richness of the environment also contributes to the subsequent behavioral and learning outcome of these children.

Vitamins

While there is considerable controversy over whether subtle or subclinical vitamin deficiencies can affect mental or behaviorial functioning and even contribute to the development of mental illnesses, there is little question that marked vitamin deficiency can have a major impact on mental functioning. Cases of vitamin B_{12} deficiency without the usual hematological abnormalities but with a frank psychosis are not uncommon. Less common are cases of pyridoxine deficiency contributing to the development of mental illness. Increased requirements for vitamins may contribute to the commonly observed depression associated with contraceptive hormone preparations. The elderly seem particularly susceptible to the development of vitamin deficiency disorders, including mental dysfunction. Their nutritional deficiencies may be due to a lack of access to adequate nutrition, a habitually deficient diet, altered requirements in old age, or added requirements due to illness.

Korsakov's psychosis is a particularly interesting model of a vitamin deficiency related condition. Studies suggest that alcoholics who develop Korsakov's have an abnormal sensitivity to thiamine deficiency; certain thiamine-dependent enzymes in these individuals have markedly abnormal saturation curves. One implication of these studies is that Korsakov's psychosis might be preventable if adequate thiamine intake were maintained in alcoholics.

Minerals

The minerals zinc, copper manganese, cobalt, iron, and cadmium are important cofactors in brain functioning. Hurley's studies in 1981 of zinc deficiency relate directly to central nervous system development. Experimental animals made zinc or manganese deficient during pregnancy produce a high proportion of offspring with structural abnormalities of the nervous system. The relevance of these studies to human mental illness is unclear, since these levels of deficiency have not been studied in humans.

Copper is a cofactor in the rate-limiting step in the synthesis of catecholamine neurotransmitter substances. Alterations in copper levels thus might be expected to have some effect on behavior, and one form of schizophrenia has been attributed to elevated levels of copper. It is unlikely, however, that alterations in diet significantly affect brain copper levels.

The most common dietary mineral deficiency is of iron. Children with relatively minimal iron deficiency—levels which do not produce anemia—perform less well on standardized neuropsychological tasks compared with non-iron-deficient children.

Dietary neurotransmitter precursors

Observations that dietary precursors of brain neurotransmitter substances can influence the levels of those substances and the brain functions mediated by them have led to strategies for the treatment of mental disorders via dietary manipulation. Addition of precursors to the diet can affect the levels of indoleamines, catecholamines, acetylcholine, glycine, and histamine.

Adverse reactions to foods and psychopathology

It has been proposed that some susceptible individuals with sensitivities to particular foods may develop a wide variety of mental dysfunctions as a result of ingestion of that food. Depression, anxiety, tension, irritability, hyperkinesis, learning disabilities, schizophrenia, and difficulty controlling aggressive impulses have been attributed to these adverse food reactions. While there is an extensive anecdotal literature addressing this issue, the nature of the relationship between food ingestion and behavioral dysfunction is unclear. No adequate mechanisms have emerged to explain these observations. Similarly, the widely held notion that food additives and sugar can produce abnormal behavior in children lacks a strong scientific underpinning. It is, however, noteworthy that in rigorous studies of additives and sugar, some individuals appear to show behavioral reactions to these substances.

A variant of this food sensitivity and behavioral dysfunction hypothesis is the notion that wheat gluten proteins can contribute to the expression of schizophrenic symptomatology. Some schizophrenic subjects placed on a gluten-free diet appear to show more rapid improvement than those receiving a regular diet. Attempts to replicate this finding with other schizophrenic patients have not, however, indicated a major role for a wheat gluten/schizophrenia relationship, although some individuals with schizophrenic symptoms may be gluten sensitive.

Further reading

Growdon JH (1979): Neurotransmitter precursors in the diet: Their use in the treatment of brain diseases. In: *Nutrition and the Brain* Wurtman R, Wurtman J, ed. 3:117–181

Pollitt E, Leibel R (1976): Iron deficiency and behavior. *Pediatrics* 88:372–381

Stein Z, Susser M., (1976): Maternal starvation and birth defects. In *Birth Defects: Risks and Consequences*. Hook, Ernest B., (ed) Academic Press, New York pp 205–220

Weiss B (1982): Food additives and environmental chemicals as sources of childhood behavior disorders. *J Am Acad Child Psychiat* 21:144–152

Winick M, Meyer KK, Harris RC (1975): Malnutrition and environmental enrichment by early adoption. *Science* 190:1173–1175

Mental Retardation

Hugo W. Moser

The American Association on Mental Deficiency recommends that mental retardation be defined as "significantly subaverage general intellectual functioning resulting in or associated with concurrent impairments in adaptive behavior and manifested during the developmental period." While at first glance this definition may appear to be self-evident and innocuous, it was arrived at only after a great deal of discussion and takes into account biological, psychological, social and even economic factors. Particularly important points are the insistence that general intellectual function must be significantly subaverage. Intellectual function here is operationally defined as the results obtained by assessment with one or more individually administered standardized general intelligence tests. With these tests the average IQ is set at 100, and the standard deviation has been found to be 15. Significantly subaverage intellectual function is defined as that which is more than 2 standard deviations below the mean IQ, that is, 70 or below. This upper limit of 70 is meant to be a guideline, and depending upon circumstances it can be extended to 75. Another significant aspect of the definition is that for a person to be classified as mentally retarded, the significantly subaverage intellectual function must be accompanied by impairment in adaptive behavior. This requirement is added in order to avoid inappropriate reliance on, or tyranny of, an IQ number. It is well known that IQ tests may underrepresent a person's potential, such as when a North American IQ test is administered to a person who has been educated in another language or culture. Impairment in adaptive behavior is defined by Grossman as "significant limitations in an individual's effectiveness in meeting the standards of maturation, learning, personal independence and/or social responsibility that are expected for his or her age level and cultural group, as determined by clinical assessment and, usually standardized scales." The developmental period is defined as the period of time between conception and the 18th birthday. The definition of mental retardation is restrictive because we have learned that the inappropriate label of mental retardation may have serious consequences. Simon Olshansky has written: "To schools the category of mental retardation is a way of classifying some students with learning problems; to the person so labeled the categorization is an attack from which recovery is rarely complete."

The two-group approach

Under this current definition between 2% and 3% of the general population would be classified as mentally retarded. It has been found useful to subdivide the mentally retarded population into two groups, the characteristics of which are outlined in Table 1.

Group 1 includes 75–85% of the total. Here the intellectual impairment tends to be relatively mild, and sociocultural deprivation and economic status are significant factors. The causes responsible for the suboptimal function in this first group are a matter of controversy. What is the relative importance of genetic and environmental factors? This question is difficult to answer. First the group is almost certainly heterogenous. Second, while there is no demonstrable abnormality of brain structure or function, techniques used to demonstrate such abnormalities are still primitive and could fail to detect them. Third, twin studies and other epidemiological studies have shown, not surprisingly, that intelligence is influenced strongly by genetic factors. In spite of these considerations, recent intervention studies have shown that early educational and social intervention techniques can do much to prevent the suboptimal cognitive development of children at risk of group 1 sociocultural mental retardation. This finding is highly encouraging and has practical and theoretical significance. It suggests that it is possible to interrupt the vicious cycle in which the disadvantaged parents who function at a suboptimal cognitive level have children who again function suboptimally. The mechanism by which these techniques of early intervention provide favorable effects is not clear. The implications in terms of human values and national policy are significant.

Table 1. Two Groups of Mentally Retarded

Characteristic	Group 1	Group 2
Names	physiological subcultural culturofamilial familial	organic pathological
Estimated incidence	20–30 per 1,000	3 per 1,000
Most frequent IQ score	50–70	less than 50
Most common age of ascertainment	during school years	preschool
Apparent change of prevalence with age	apparently diminishes after school years	no change
Demonstrable brain abnormality	no	yes
Relationship to socioeconomic status	more common among socioeconomically deprived or disrupted families	none or slight

Group 2, or pathological mental retardation, is somewhat more readily approachable by techniques familiar to neuroscientists. It is convenient to set an IQ of 50 as the dividing line between the two groups. It has been found that the prevalence of this form of mental retardation (group 2) is relatively constant throughout the world: It varies between 3 and 4 per 1000. Over 90% of persons with an IQ of less than 50 have demonstrable abnormalities of brain structure or function. The IQ 50 dividing line is not absolute: There are mentally retarded persons with demonstrable brain pathology whose IQ is higher than 50, and extreme cultural deprivation alone can lead to apparent IQ levels of less than 50.

Causes of severe mental retardation

The causes of severe mental retardation have been the subject of intensive studies (Table 2). They have been shown to be prenatal in 55% of cases and chromosomal in 29%. The term chromosomal disorder refers to a group of disorders in which there is an abnormality of the number or appearance of chromosomes which can be detected with the light microscope. Down's syndrome is the most common chromosomal disorder associated with mental retardation, followed by the fragile X-syndrome. In this X-linked disorder, there is a fragile site on the distal portion of the X chromosome, an abnormality which is brought about when cells are grown in a medium which contains low levels of folic acid or chemicals such as 5-fluorodeoxyuridine. The single mutant genes include inborn errors of metabolism and in the aggregate account for 5% of severe mental retardation. Included here are several hundred separate entities. The most common are phenylketonuria, congenital hypothyroidism, and the urea cycle disorders. Intrauterine infections, such as rubella, cytomegalovirus, and herpes simplex account for 7%; perinatal events for 12%; and postnatal for 11%. Truly remarkable advances in neonatal care have reduced greatly the percentage of surviving low birthweight infants who suffer brain damage. By the same token the total number of low birthweight infants has increased, so that the number of low birthweight infants who survive with brain damage has remained relatively constant.

Table 2. Causes of Severe Mental Retardation

	% of Total
Prenatal	
Genetic chromosomal (Down's, fragile X, sex chromosome disorders)	29
Single mutant genes (phenylketonuria, hypothyroidism, and many others)	5
Multiple congenital anomaly and mental retardation syndromes	12
Acquired; fetal alcohol syndrome, intrauterine infections (rubella, cytomegalus virus, etc.)	15
Perinatal	
Fetal anoxia, asphyxia,	12
infections, etc.	3
Postnatal	11
Psychosis	1
Unable to determine cause	18

It is encouraging to note that several forms of severe mental retardation can now be prevented or treated. Greatest success has been achieved with the immunization programs against rubella and measles—the mass screening of newborns for phenylketonuria and hypothyroidism and the prompt and successful institution of dietary therapy for the former and thyroid replacement for the latter. Other notable successes are immunization programs for Rh-negative mothers, which prevent the previously dreaded disorder erythroblastosis fetalis. Study of fetal cells or amniotic fluid obtained by aminocentesis can prenatally diagnose Down's syndrome, malformations such as anencephaly or spina bifida, and more than 100 inborn errors of metabolism. These and other approaches have the potential of bringing about a significant reduction in the incidence and extent of disability of severe mental retardation. Some of these treatable causes are highlighted in Table 3. In addition, recent and future advances in molecular biology and developmental neurobiology make it likely that additional causes of severe mental retardation can be understood and prevented or treated.

Table 3. Causes of Severe Mental Retardation

Potential for Prevention Condition	Incidence	Approximate Estimated or Treatment
Down's syndrome	1:1000	Partial reduction (about 15–30%) through awareness of greater risk for mothers age 35 or above and prenatal studies for those at risk.
Fragile X	?approx. 1:1000 in males	Potential for prenatal diagnosis in pregnancies known to be at risk.
Neural tube defects	1:750 mental retardation in about two-thirds	Alpha-fetoprotein screening in pregnancies at risk can bring about 15% reduction. Potential for screening maternal plasma alpha fetoprotein to permit better identification of those pregnancies that are at risk.
Fetal alcohol syndrome	1:700	Potential for prevention through public education.
Congenital hypothyroidism	1:4500	Mental retardation can be prevented completely by mass screening of newborns and thyroid hormone administration to affected infants.
Phenylketonuria	1:14,000	Mental retardation can be prevented completely by mass screening of newborns and administration of special diet.
Intrauterine viral infections	1:5000	Rubella preventable nearly 100% through mass immunization programs.
Mental retardation due to perinatal anoxia or asphyxia particularly in low birthweight infants	1:2500	Two preventive approaches: reduce incidence of premature births, particularly in adolescent mothers; further improvements in neonatal care—regional centers and neonatal intensive care units.

Further reading

Crome L, Stern J (1972): *Pathology of Mental Retardation,* 2nd ed. Edinburgh: Churchill Livingstone

Grossman HJ (1983): *Classification in Mental Retardation.* Washington DC: American Association on Mental Deficiency

Hagberg B, Kyllerman M (1983): Epidemiology of mental retardation: A Swedish survey. *Brain Dev* 5:441–449

Olshansky S (1970): Work and the retarded. In: *Diminished People,* Bernstein NR, ed. Boston: Little, Brown, pp 29–46

Moser HW (1977): Mental retardation. In: *Horizons of Health,* Wechsler H, Gurin J, Cahill GF Jr, eds. Cambridge: Harvard University Press

Moser HW, Ramey CT, Leonard CO (1983): Mental retardation. In: *Principles and Practice of Medical Genetics,* Emery AL, Rimoin DL, eds. Edinburgh: Churchill Livingstone, pp 352–366

Monoamine Oxidase (MAO) Inhibitors in Psychiatric Therapy

Dennis L. Murphy

Iproniazid and several other hydrazine inhibitors of the oxidative deamination of brain catecholamines and indoleamines were the first drugs shown to be effective in the treatment of major depressive disorders. Although iproniazid itself proved to be hepatotoxic and is no longer used clinically, and the tricyclic antidepressants came to be regarded as safer and more broadly effective drugs, other monoamine oxidase (MAO) inhibitors continue to be widely used, especially in tricyclic-resistant depressed patients.

To meet Food and Drug Administration requirements, the therapeutic efficacy of several of the standardly used MAO inhibitors (tranylcypromine, isocarboxazid, and phenelzine) was reevaluated in the early 1980s, and antidepressant effects comparable to those of the tricyclics were demonstrated. In addition, several newer types of MAO inhibitors have been developed, some with definite therapeutic potential which are also of interest as tools to explore the importance of different monoamine systems in brain function and in the mechanism of action of antidepressants.

Most of these drugs are phenylethylamine derivatives, producing irreversible inhibition by covalently binding to the FAD-containing active site of MAO via such substituent groups as a hydrazine (-N-N-, e.g., phenelzine, isocarboxazid), cyclopropylamine (-C—C-, e.g., tranylcypramine), or acetylenic
\ /
C
group (—C≡C—, e.g., pargyline, clorgyline, deprenyl). Because brain MAO has a half-life of 12 days, these drugs have a very long biological duration of action, with consequences for clinical use that include some difficulties in titration of dosage, lack of drug plasma level monitoring utility, and prolonged toxicity, when it occurs. A series of new, reversible MAO inhibitors, including moclobemide, CGP-11305A, cimoxatone, and amiflamine, are under intensive early clinical study to determine if equal efficacy, improved dosage control, and fewer side effects might be found.

The acetylenic inhibitors, clorgyline and deprenyl, are of special interest because they act selectively to inhibit the two major subforms of the enzyme, MAO type A and MAO type B, respectively. In rodent and human brain, clorgyline inhibits the deamination of serotonin, a substrate for MAO-A, at tissue concentrations over 1,000-fold lower than those needed to inhibit the deamination of phenylethylamine, an MAO-B substrate. Conversely, deprenyl selectively inhibits the deamination of phenylethylamine and other subtrates of MAO-B, a difference that has been put to clinical use in the treatment of Parkinson's disease. Dopamine is predominately an MAO-B substrate in humans, and deprenyl appears to potentiate the actions of L-dopa in alleviating some symptoms of this disease without potentiating noradrenergically mediated cardiovascular effects, a complication found when nonselective MAO inhibitors are combined with L-dopa. Clorgyline appears to be a more effective antidepressant in severely depressed patients than deprenyl or the partially selective MAO-B inhibitor pargyline, suggesting that the major antidepressant effects of the MAO inhibitors are more likely to be mediated via central noradrenergic or serotonergic neurotransmitter systems than by changes in phenylethylamine or dopamine metabolism.

A cascade of cellular events appears to be involved in the mechanism of action of these drugs. Acutely, especially with high MAO inhibitor doses, an increase in the intracellular concentrations of brain monoamines occurs. Not only are neuron-specific amines increased (e.g., norepinephrine levels in noradrenergic cell bodies and presynaptic processes) but other amines, whose concentrations are normally very low in brain (e.g., tryptamine, phenylethylamine, and epinephrine) are even more markedly elevated. Vesicular concentrations of amines are increased, as are cytoplasmic concentrations, which normally are very low. A slowing of neuronal firing rates has been demonstrated in both serotonin-containing neurons in the median raphe and in norepinephrine-containing neurons in the locus ceruleus. Further adaptive consequences of MAO inhibitors include a reduction in amine synthesis, which has been most clearly demonstrated within the noradrenergic system. After more chronic treatment, reductions develop in beta-adrenoreceptor numbers and in beta-receptor functional activity—as measured by norepinephrine stimulated cyclic adenosine monophosphate formation—as well as in alpha$_2$-adrenergic binding sites and serotonin binding sites. These receptor changes are generally similar to those found after tricyclic antidepressant treatment in animals, and have been hypothesized to be involved in the therapeutic actions of antidepressant drugs.

Some physiological consequences of these cellular events that are found in humans include an essentially complete suppression of rapid eye movement sleep, a change produced by clorgyline but not deprenyl, and hence attributable to MAO-A inhibition. Decreased blood pressure, associated with decreased plasma concentrations of norepinephrine and its metabolite, 3-methoxy, 4-hydroxyphenylglycol, suggests a suppression of central sympathetic outflow. Not unexpectedly, orthostatic hypotension is a side effect encountered with these drugs. At the same time, however, increased cellular concentrations of norepinephrine may be released in the periphery upon the ingestion of foodstuffs containing tyramine or similar indirectly acting sympathomimetic drugs, producing the most dangerous side effect and liability of MAO inhibitor treatment, a hypertensive crisis. Dangerous interactions also occur centrally if, during chronic MAO inhibitor treatment, tricyclic antidepressants or some synthetic narcotics like meperidine are administered.

Nonetheless, large-scale studies involving several thousand patients have shown a very low incidence of major complications when dietary and drug interaction precautions are provided to patients prior to MAO inhibitor treatment. In fact, with the current increased interest in depressed patient subgroups and the improved diagnostic delineation of psychiatric syndromes, clinical trials of MAO inhibitors have been on the increase in the 1980s. The older literature suggested that a heterogeneous, ill-defined subgroup of atypical depressives might respond preferentially to MAO inhibitors. More recently, patients meeting diagnostic criteria for panic disorder, agoraphobia, bulimia—disorders frequently associated with depression—as well as some more clearly identified subgroups of major depressive disorder patients have been found to respond to one or another MAO inhibitor in randomized, double-blind trials. Some evidence suggests not only that the substrate-selective inhibitors like clorgyline and deprenyl may have different clinical effects, but also that other MAO inhibitors among both the irreversible and the newer reversible drugs may differ in their spectrum of efficacy, a point for further study.

Further reading

Beckmann H, Riederer P, eds (1983): *Monoamine Oxidase and Its Selective Inhibitors: New Concepts in Therapy and Research.* Basel: Karger

Murphy DL, Garrick NA, Cohen RM (1983): Monoamine oxidase inhibitors and monoamine oxidase: Biochemical and physiological aspects relevant to human psychopharmacology. In: *Drugs in Psychiatry, Vol 1, Antidepressants,* Burrows JD, Norman TR, Davies E, eds. Amsterdam: Elsevier/North Holland Biomedical Press

Squires RF (1978): Monoamine oxidase inhibitors: Animal pharmacology. In: *Handbook of Psychopharmacology, Vol 14,* Iverson LL, Iversen SD, Snyder SH, eds. New York: Plenum Press

Tipton KF, Dostert P, Strolin Benedetti M, eds (1984): *Monoamine Oxidase and Disease.* London: Academic Press

Mood Disorders

Alan J. Gelenberg

Emotional reactions are part of our biological heritage as social mammals. For us, as for our prehuman and preprimate forbears, emotions appear to have survival value—for species, as well as for individuals. Particularly in the vulnerable years of infancy (an especially long period in *Homo sapiens*), the behavioral expression of emotions, with their frequent accompaniment of interpersonal responses, enhances the likelihood of biological survival. For example, a newborn experiences the loss of physical support: his eyes open wide, his heart races, he cries; the observer interprets the child's experience as anxiety, feels his own discomfort, moves to readjust the infant to provide support. Or, a baby's mother goes away. After an initial bout of protest, the child lapses into a period of apathy and despair ("anaclitic depression"). This will engender in most adults a desire to provide emotional support for the child, such as physical support was rendered in the earlier instance. And similarly, even in adulthood we communicate both physiological and psychological needs through our emotional expressions, needs usually meetable and met by kith and kin.

Our feelings run the gamut from anger, sadness, and guilt to joy. In a normal individual these reactions, or affects as psychiatrists call them, will be appropriate in degree and kind to the circumstances that evoke them. For example, we expect profound sadness in reaction to the death of a loved one; elation or indifference would be cause for surprise. On the other hand, prolonged dejection in reaction to a relatively trivial loss would be equally perplexing.

In what has become a well-worn psychiatric simile, what weather is to climate, affect is to mood. One may observe a relatively full range of an individual's affective responses during a serious and involved interview. But to discern his mood—the undertone of his state of mind over many days, weeks, or longer—one must combine direct observations, historical questions, and interviews with people who know him. The normal individual, absent extraordinary conditions, carries a fairly neutral mood, from which he reacts to pleasant or unpleasant events with the appropriate emotion.

Our understanding of the biology of normal emotions and mood remains primitive. Probably the hypothalamus, the limbic system, and portions of the cortex are involved in the subjective experience and behavioral manifestations of different emotions. Limited experimental evidence suggests a possible role for the neurotransmitters norepinephrine and serotonin and for the endogenous opioid systems in the mediation of affective states. But it remains for future investigations to map out the definitive neuroanatomy, neurophysiology, and neurochemistry of mood to bring modern scientific understanding to profound and ancient experiences.

Pathology of mood

Since Biblical times, and doubtless before, some troubled souls have been tortured by bouts of depressed mood. For weeks, for months, even for years they might lie in apathy, or languish in states of hellish agitation and torment. The depressed individual could be plagued with self-doubt, guilt, and recriminations. Probably he would lose interest in formerly pleasurable activities, often including eating and sex. Sleep, weight, and other bodily functions would typically be altered. Such a patient might lose touch with reality, experiencing unvalidated sensations (hallucinations) or adopting false, fixed beliefs (delusions). Not infrequently, the sufferer would attempt or actually succeed in committing suicide. Sometimes the depression would be preceded by an unidentifiable loss or emotional trauma, but at other times it might occur spontaneously, or possibly as part of some rhythmic cycle (perhaps associated with a change of seasons).

Less common, but perhaps more impressive to observers, are episodes of pathological elation that occur in some individuals afflicted alternately with bouts of depression. A patient in the throes of mania typically manifests a euphoric mood, with unreasonable confidence in his own abilities and unbridled optimism about the future. (At times the euphoria switches gears into irritability and possibly a pathological suspiciousness—paranoia). A manic patient is likely to monopolize conversations, tie up the telephone, run up outrageous bills, and expend almost limitless energy with little apparent need to sleep.

Perhaps between 5% and 15% of the population can expect at least one episode in their lives of what is currently termed major depression. Diagnosed on the basis of observations and historical information, a major depressive episode consists of a moderate to severe lowering of mood for at least a period of weeks accompanied by a requisite number of associated symptoms, including disturbances of sleep, appetite, and energy, and mental changes such as a preoccupation with death or suicide. For at least half those who experience a major depressive episode, similar episodes will recur in the future, at times frequently and to a debilitating degree, and in some taking on a chronic course without periods of remission.

When episodes of depression alternate with periods of mania, psychiatrists diagnose the bipolar form of mood disorder. The lifetime prevalence of this illness is probably between 0.5% and 1%, although its social impact in the marketplace and on families probably is greater on a case-by-case basis than that of recurrent depressions alone.

In some individuals, episodes of depression appear clearly related to life events, such as the loss of a job or the breakup of a relationship. At times, the link between events and emotional reaction appears "neurotic," as when a man reacts to success by becoming depressed. At other times, episodes of depression or elation appear entrained to natural rhythms such as daily, monthly, or annual cycles. In still other patients, mood swings come and go with little predictability and few apparent correlates.

For some, depression becomes almost a way of life. But where "character" or "personality" begins and illness leaves off is scientifically undefined. Perhaps some individuals adopt depressive behaviors to satisfy ill-defined intrapsychic needs or to evoke desired behaviors in those around them. Alternatively, chronic depression could reflect a lifestyle adapted to an illness of biological dysregulation.

Disturbances of mood tend to run in families, yet in epidemiological surveys it is hard to disentangle the contributions of nature from those of nurture. However, studies of biological markers, of dizygotic versus monozygotic twins, and of adoptees and their families suggest that at least some components of some forms of mood disorder are genetically transmissible. The best evidence exists for bipolar illness, although many relatives of bipolar patients have a unipolar (recurrent) depressive disorder. It is interesting to speculate on what actually might be inherited: an illness, some form of regulatory disturbance, a deficiency in some "pleasure center," or a handicapped ability in an indeterminate interpersonal function or ability to calm, soothe, or derive pleasure in oneself.

Much interest in the biology of mood disorders has followed in the wake of discovery of antidepressant drugs. Yet despite a plethora of hypotheses, our understanding of the pathology of emotion, much as that of its normal physiology, remains rudimentary. Early theories postulated deficits in cerebral levels of norepinephrine and serotonin, and while these two neurotransmitters still enjoy the lion's share of biological attention, more recent theories have been focused on altered sensitivity of their receptors, both pre and postsynaptic. However, the direction of this putative altered sensitivity—whether increased, decreased, or unstable—remains undetermined. Other neurotransmitter systems, such as that involving acetylcholine, also might be involved, and it would not be surprising to find the ultimate answer involving interrelationships among a number of systems. Moreover, different patients may suffer from different biochemical lesions.

Psychiatrists have long envied specialists in infectious disease, who practice in the ideal "medical model." These fortunate physicians can treat readily defined illnesses caused by identified pathogens. The ability to isolate bacteriological and viral causes has freed them from a need to focus on the patterns of fever and other clinical signs. But although psychiatrists have many notions about biological subtypes of major depression and their differing likelihood of response to pharmacologic treatments, we still lack definitive biological tests that are essential for defining and scientifically treating disease states. In recent years much effort has gone into the development of such tests, such as those based on probing neuroendocrine responses and on the excretion of endogenous amine metabolites. Despite great expenditure of time and funds, however, rigorous biological diagnosis of major mood disorders remains elusive.

The personal, social, and financial costs of mood disorders are tremendous: suicides, "accidents," hospitalizations, wrecked or compromised careers, impaired parenting, broken marriages, squandered capital. Thus society has a moral as well as a pragmatic stake in identifying cases, facilitating treatment, and promoting research toward prevention and cure.

Treating depression and mania

In the early part of this century, biological therapies of mood disturbances were very limited. For the agitation, anxiety, or insomnia of depression, as for the elation of mania, a doctor could offer only sedatives—such as bromides, barbiturates, or chloral hydrate. But in addition to their limited effectiveness, these agents have a fairly narrow range of safety, with a rapid passage from sedation to cessation of breathing and other vital functions. When amphetamines and other stimulants became available, they were widely tried in depression, but their efficacy likewise is limited, and problems of tolerance and abuse further limit their clinical utility. (Stimulants have been enjoying a limited renaissance of late in some medical circles, but definitions of their role, if any, in the treatment of mood disorders must await more investigation.)

Although the role of psychotherapeutic interventions was never formally studied, most clinicians found them more useful in milder forms of depression, where patients were more amenable to talking and self-reflection. Although some practitioners believed (and continue to believe) in a role for insight-oriented verbal psychotherapy in the treatment of profound and psychotic depression and in mania, there is no rigorous evidence in support of this practice.

In the 1930s, investigators discovered the healing power of induced seizures, and electrically provoked convulsions became an accepted form of treating severe depression. (To this day, the mechanism by which electroconvulsive therapy (ECT) performs this function remains unknown, although few psychiatrists question its efficacy.) In the 1950s, two serendipitous discoveries gave rise to two chemically and pharmacologically distinct groups of antidepressant drugs: the tricyclics (named for their chemical structure) and the monoamine oxidase (MAO) inhibitors (named for their inhibition of the enzyme that catabolizes several important neurotransmitters). And only a few years earlier, Cade reported on the antimanic properties of the cation lithium. These discoveries gave rise to a renaissance in biological psychiatry, which continues today. The advent of drugs to treat mania and depression has spawned vigorous research into the pathophysiology of the disorders and the biological mechanism of the treatment; the use of drugs has also afforded relief to millions of sufferers from these painful and sometimes fatal conditions.

Antidepressant drugs bring significant relief to 70–80% of patients suffering from major depression. Most likely to respond are patients suffering from the melancholic or "endogenous" form of this disorder, in which mood is relatively autonomous (i.e., displaying little tendency to vary with the immediate environment); accompanying symptoms include early morning awakening, feeling worse earlier in the day, and loss of appetite and weight. Research continues into the possibility that some types of antidepressants, specifically the MAO inhibitors, may be particularly effective for yet other kinds of depression. Some individuals with recurrent episodes of severe depression benefit from long-term maintenance therapy with antidepressant drugs, which suppress or prevent the recurrence of future episodes.

Since the introduction of the original tricyclic and MAO inhibitor antidepressants in the 1950s, a number of similar agents have been introduced in the U.S. and other markets. These related drugs have extended physicians' options somewhat, as some patients are more likely to respond to one drug than another. Similarly, since the drugs have different side effects, the clinician is better able to tailor drug to patient. In recent years a number of new drugs have been introduced, which are relatively distinct chemically (and sometimes pharmacologically) from the original agents. However, many are associated with new forms of toxicity, which encumber their clinical utility. Furthermore, while it is generally acknowledged that none of the newly introduced drugs is superior to the old standbys, there is increasing concern among clinicians that some are not equal in efficacy to the older agents. The search continues for antidepressant drugs that are more effective and have fewer adverse effects and less toxicity.

Lithium has become a mainstay in the treatment of bipolar illness. It is effective in the treatment of acute mania, although in its more severe forms mania may require the use of an antipsychotic drug. Probably lithium's greatest utility comes in the long-term management of bipolar illness, where most patients enjoy a marked diminution in the number of episodes of mania and usually of depression as well. Lithium may also play a role in the long-term maintenance therapy of patients with only recurrent depression and no episodes of mania.

Both lithium and antidepressant drugs affect uptake of neurotransmitters in presynaptic nerve terminals and also alter sensitivity of pre and postsynaptic neuronal receptors. Despite our emerging knowledge about these effects, the actual mechanisms of action of these agents are still unclear.

ECT remains a backup therapy in patients with severe depression (or even mania) when drugs are ineffective or medically contraindicated. Much of the mythology about ECT is based on the earlier techniques employed, while the modern form is particularly safe and associated with relatively few complications. Newer forms of drug therapy are currently being investigated for the treatment of both depression and mania.

In addition to the biological treatments of mood disorders, modern investigations also have focused on interpersonal interventions. In particular, cognitive and behavioral treatments have been defined and are currently being studied as possible approaches to specific types of depression and as adjuncts to biological treatments.

As is typical in medical science, interest in understanding of disorders follow improved treatments. The last 30 years have brought major advances in the treatment of more severe forms of mood disturbance, and physicians and scientists may be hopeful that the coming decade will bring still further advances in therapies and accompanying improvement in our understanding of these life-threatening, disruptive, and extremely painful illnesses.

Further reading

Baldessarini R (1983): *Biomedical Aspects of Depression and Its Treatment.* Washington: American Psychiatric Press

Bassuk E, Schoonover S, Gelenberg A (1983): *The Practitioner's Guide to Psychoactive Drugs.* New York: Plenum Publishing Corporation

Consensus Statement on Long Term Prevention of Recurrent Mood Disorders. *Am J Psychiatry* 1985. 142(4):469–476

Gelenberg A, ed. *Biological Therapies in Psychiatry* (monthly). Littleton, Mass: PSG

Morphine

Terrance M. Egan

Morphine is the chief active ingredient of crude opium, the dried sap of the poppy *Papaver somniferum*. The drug has pronounced actions on both the central nervous system (CNS) and peripheral nervous system, and is clinically useful as an analgesic and antitussant, and in the control of diarrhea. It is also a powerful narcotic which is widely abused for its euphoric qualities, often resulting in drug addiction.

Laboratory synthesis of morphine is possible but difficult, so the drug is usually obtained from crude opium. The method of collecting opium is described in Assyrian tablets dating to the 7th century BC. The unripe seed capsule of the poppy plant is incised, causing a milky exudate to seep from the wound. This exudate, when dried and powdered, is opium. Opium contains phenanthrene and benzylisoquinoline alkaloids. The chief phenanthrene, making up about 10% of the crude opium, is (-)-morphine. Morphine was isolated from opium by Serturner, a German pharmacist's apprentice, in the early 1800s. Serturner extracted an alkaloid, which he called morphine after Morpheus, the Greek god of sleep and dreams, that retained many of the familiar pharmacologic properties of opium. The importance of Serturner's discovery was not immediately appreciated. Morphine is a base of pKa 8.5, and is therefore poorly absorbed from the gastrointestinal tract, making it ineffective when given orally. However, the isolation of morphine was followed shortly by an increase in the popularity of the use of the hypodermic needle in medicine. Parenteral administration of morphine was widely used for the control of pain during the American Civil War, leaving a large number of soldiers addicted to the drug.

The structure of morphine was determined by Gulland and Robinson (1926) (Fig. 1A). (-)-Morphine is a phenanthrene alkaloid; however, the drug is probably best classified as an *N*-methylpiperidine compound with bulky ring substituents. This distinction is born out of the study of the relationship of the structure of the molecule and its ability to produce analgesia. Modification of the phenanthrene nucleus produces only quantitative changes in analgesic activity. For example, meperidine lacks most of the phenanthrene ring structure yet retains analgesic action. In contrast, modification of the piperidine ring results in both quantitative and qualitative changes, and is generally thought to be essential for activity (however, close examination of the structure of methadone, which lacks an intact piperidine ring, shows that there are exceptions to this rule). In simplified form, there are three parts of the morphine molecule required for activity: (1) a phenolic ring, (2) a quaternary carbon (C13) to which the phenolic ring is attached, and (3) a nitrogen atom separated from the quaternary carbon by a two-carbon chain (C15, C16) (once again, there are exceptions). In morphine, a methyl group is attached to the nitrogen. When an allyl group is substituted for this methyl, the resulting compound, nalorphine, shows both agonist and antagonist activities. Likewise, substitution of allyl for methyl on the nitrogen of the semisynthetic morphine derivative, oxymorphone, produces naloxone (Fig. 1B), a pure narcotic antagonist.

The earliest record of the medicinal use of morphine (as opium) is Sumerian (300–400 BC), probably in the treatment of diarrhea. Morphine inhibits peristalsis, and causes spasm of the smooth muscle of the intestine, especially at the sphincters. These actions result from a direct action of the drug on the nerves of the gastrointestinal tract and probably do not require the involvement of the CNS. However, most of the other desirable and undesirable actions of morphine do involve the CNS. The most popular medicinal use of the drug is as an analgesic in the control of dull intermittent pain. Its principal action is to dissociate pain from its implication, that is, to separate the perception of pain from the emotional reaction to it. The analgesia is selective; the drug is not effective against acute, intermittent pain, and other sensory modalities are spared. Morphine is thought to act at several different levels of the CNS to produce this effect. Morphine is also effective in suppressing cough; however, the related opiate alkaloid, codeine (another natural substituent of opium), is usually used for this purpose.

Expression of the actions of morphine results when the drug combines with specific proteins contained within cell membranes of target tissues. These proteins constitute several classes of ligand-specific opioid receptors. The existence of these receptors was first predicted from the early work of Martin (1967), and was subsequently verified by several laboratories. At present, there are at least three separate subclasses of opioid receptors, denoted by the Greek letters μ, δ, and κ. Of these three, morphine has the highest affinity for the μ-receptor, and, in functional studies, appears to be a good μ-selective agonist. The presence of opioid receptors suggests that endogenous ligands for these receptors exists. A number of peptide hormones have been isolated. These peptides bind opioid receptors with high affinity, and in so doing, produce many of the actions of morphine.

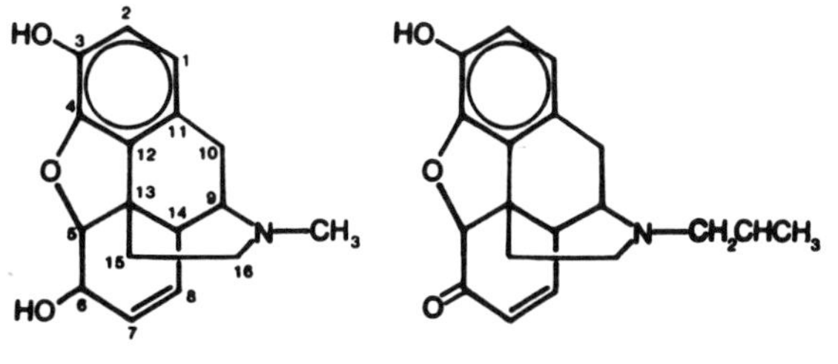

Figure 1. A. Morphine. B. Naloxone.

The relative importance of the events that follow the combination of morphine and receptor are still disputed. At least three such events are well documented. First, several lines of evidence suggest that morphine causes an inhibition of adenylate cyclase. Second, in both brain and gut neurons, morphine causes naloxone-reversible hyperpolarizations by increasing the membrane conductance to potassium ion. Third, morphine causes a decreased transmitter release. That there is a causal relationship between combinations of these effects seems likely, but is unproved.

Further reading

Akil H, Watson SJ, Young E, Lewis ME, Khachaturian H, Walker JM (1984): Endogenous opioids: Biology and function. *Annu Rev Neurosci* 7:223–255

Johnson MR, Milne GM (1981): Analgetics. In: *Burger's Medicinal Chemistry*, 4th ed, Wolf ME, ed. New York: John Wiley

Musto, DF (1973) *The American Disease: Origins of Narcotic Control*. New Haven: Yale University Press

Neuroleptic Drugs

Solomon H. Snyder

Neuroleptic drugs are agents of diverse chemical structure but have a single major clinical action, the relief of psychotic symptoms, especially schizophrenic ones. These drugs are also referred to as major tranquilizers and antipsychotics. The term neuroleptic is most commonly used, since it is more specific. By contrast, the terms antipsychotic and tranquilizer are employed in different contexts with diverse clinical conditions and can apply in principle to many classes of drugs.

The first neuroleptic drug was chlorpromazine. Since chlorpromazine comes from the phenothiazine chemical class and since many other phenothiazine drugs have been employed as neuroleptics, the term phenothiazine has often been used interchangeably with neuroleptic. However, other chemical classes, especially the butyrophenones and the thioxanthenes, have the same pharmacologic and therapeutic properties as the phenothiazines.

Understanding the pharmacologic actions of neuroleptics in animals and humans requires an appreciation of the history of these drugs in clinical medicine. Chlorpromazine was synthesized by the Rhone-Poulenc Drug Company in France as an antihistamine related to the antihistamine promethazine, which also was of the phenothiazine chemical class. The neurosurgeon Henri Laborit evaluated promethazine as a possible adjunct to preoperative anesthesia and was impressed with its ability to quiet patients without putting them to sleep. He examined chlorpromazine as an alternative which was even more sedating but still did not render patients unconscious. These properties suggested to Laborit that chlorpromazine might be useful in psychiatric patients to relieve their agitation without interfering with consciousness. The psychiatrists Jean Delay and Pierre Deniker examined chlorpromazine for these indications and were quite impressed with its ability to quiet hyperactive patients. They found beneficial effects especially in manics and schizophrenics. Though chlorpromazine was sedating, Delay and Deniker felt that they could detect definite alleviation of psychotic thinking. Over the next several years numerous psychiatrists confirmed the drug's ability to diminish psychotic behavior and especially to influence in a favorable fashion schizophrenic thinking disorders.

Clinical actions

It is now generally accepted that the major utility of the neuroleptics lies in the treatment of schizophrenia. Neuroleptics are the only drugs which reduce psychotic symptoms of large numbers of schizophrenics to a level where hospitalization is no longer required. In most patients neuroleptics do not "cure" schizophrenic symptoms. However, in a substantial number long-lasting and extensive remissions are possible. The inability of neuroleptics or any other drugs to abolish the schizophrenic disease process is apparent from the requirement for maintenance neuroleptic therapy. Numerous studies have established that patients in remission will experience a relapse of symptomatology if neuroleptic treatment is stopped. This raises the issue of the duration of maintenance neuroleptic therapy for schizophrenics in remission. It is not clear whether lifelong therapy is required or whether drug administration can be gradually reduced without the risk of serious relapse.

Neuroleptics reduce the global severity of the schizophrenic illness. However, detailed studies of specific psychological symptoms suggest that neuroleptics act upon certain symptoms more than others. Schizophrenic symptoms have been subdivided by some psychiatrists into positive and negative symptoms. The positive symptoms primarily reflect the florid, acute psychotic symptoms such as hallucinations and delusions. Neuroleptics are particularly effective in relieving these symptoms, including the typical schizophrenic disorder in thinking. Negative symptoms refer to the affective withdrawal and autistic behavior manifested by many schizophrenics and are most marked in chronic schizophrenic patients. Even in the absence of overt psychotic symptomatology such patients appear to be emotionally withdrawn from their environment. Most neuroleptics are less effective in relieving negative than positive symptoms, but one chemical class of neuroleptics, the diphenylbutylpiperidines, appears to be substantially more effective than other neuroleptics in relieving negative symptoms of schizophrenia. The diphenylbutylpiperidines are comparable to other neuroleptics in reversing positive symptoms.

Neuroleptics are also effective in relieving acute manic symptoms. However, it is not clear to what extent they act selectively upon the psychotic symptomotology rather than merely calming the patient. Often, manic patients treated with neuroleptics appear to continue the manic pressure of ideas and of speech but are slowed by the chemical straitjacket of the drug.

Neuroleptic drugs have been employed in neurotic patients for their sedating and calming actions. However, neuroleptics do not relieve anxiety in a selective fashion as do benzodiazepine drugs such as diazepam (Valium). Moreover, neuroleptics can produce serious side effects, some of which are irreversible. Accordingly, proper clinical practice restricts the use of neuroleptics to psychotic patients.

A substantial number of different neuroleptic drugs are marketed in the United States and an even greater number are marketed in Europe and other countries of the world. All the major commercially marketed neuroleptics have similar antipsychotic effects. They differ principally in their absolute potency and in their spectrum of side effects. Some of the more commonly employed agents of the phenothiazine class include chlorpromazine (Thorazine), thioridazine (Mellaril), perphenazine (Trilafon), prochlorperazine (Compazine), trifluoperazine (Stelazine), and fluphenazine (Prolixin). The thioxanthene chemical class of drugs chemically resembles the phenothiazines. Extensively used thioxanthenes include thiothixene

(Navane) and flupentixol. The butyrophenones differ markedly in their chemical structures from the phenothiazines or the thioxanthenes. However, they exert essentially the same spectrum of pharmacologic actions. Haloperidol (Haldol) is the most widely used butyrophenone neuroleptic and the only one commercially marketed in 1984 in the United States. Diphenylbutylpiperidines are chemically derived from the butyrophenones and include agents such as pimozide, fluspirilene, and penfluridol.

Side effects

For many drugs side effects are regarded as unimportant but troublesome actions of drugs. For the neuroleptics side effects are an essential part of the action of all the currently employed neuroleptics and provide insight into the mechanism of their therapeutic effectiveness. Moreover, the side effects elicited by neuroleptics are often severe and limit the clinical use of the drugs. Since the efficacies of the numerous commercially marketed neuroleptics are similar, a physician typically selects a particular neuroleptic for a given patient based on the sensitivity of the patient to side effects.

The neurological side effects provoked by neuroleptic drugs account for the designation "neuroleptic." These side effects involve the extrapyramidal motor system and often resemble the symptoms of Parkinson's disease. In their first studies of chlorpromazine Delay and Deniker observed that, as they increased doses to levels which produced therapeutic effects, they almost invariably observed Parkinson-like side effects. Accordingly, they concluded that the neurological actions of the drugs were related in some way to their therapeutic actions. The designation neuroleptic derives from Greek words meaning "to clasp the neuron" and implies a relationship of neurological to therapeutic effects. Neuroleptic drugs often elicit typical Parkinsonian symptoms essentially indistinguishable from the symptoms of patients with idiopathic Parkinson's disease. Such manifestations include rigidity, difficulty in moving, and tremor. Other neurological effects also occur with neuroleptic drugs and appear to reflect influences upon the same extrapyramidal motor areas of the brain. Acute dyskinesias may be manifest with abrupt distortions of the neck, designated "torticollis," or with abnormal movements of the eyes, referred to as "oculogyric crises." A more frequent but subtler, extrapyramidal motor abnormality produced by neuroleptics is designated "akithisia." Patients with akithisia have difficulty sitting still. They subjectively report a muscular itchiness and constantly pace up and down the room in efforts to relieve the peculiar feelings in their muscles. A physician should be alert to akithisia, since it is frequently mistaken for agitation with medical personnel prescribing increasing doses of the drug, which only makes the akithisia worse. The neurological side effects of neuroleptics occur most frequently with the butyrophenones and with the phenothiazines which are most potent on a milligram basis, such as fluphenazine.

Tardive dyskinesia is one of the most serious neurological side effects of neuroleptics. Symptomatically it looks like the opposite of the acute parkinsonian side effects. While parkinsonian symptoms include rigidity and a paucity of movement, patients with tardive dyskinesia manifest excess motor activity, especially in the muscles of the tongue and face. The fact that the symptoms of tardive dyskinesia are the opposite of the acute parkinsonian effects of neuroleptics suggests that tardive dyskinesia involves a compensation to the acute actions of neuroleptics. Tardive dyskinesia occurs predominantly in patients treated for long periods of time with large doses of neuroleptics. It has been suggested that such long-term treatment with the drugs provokes a neuronal attempt to compensate for the acute effects of the drugs.

Sedation and orthostatic hypotension occur frequently with some neuroleptics. Orthostatic hypotension refers to a fall in blood pressure when a patient stands up and is often associated with fainting spells. Sedation and hypotension occur more with phenothiazine neuroleptics, especially those which are least potent on a milligram basis, such as chlorpromazine and thioridazine. Among various neuroleptic drugs sedation and hypotension tend to go hand in hand so that most researchers feel that these two symptoms are mediated by the same or similar mechanisms. Since sedation-hypotension and neurological side effects are the two major classes of side effects, psychiatrists select neuroleptics for their patients based on sensitivity to these particular actions. The neuroleptics most prone to produce sedation and hypotension are least likely to produce extrapyramidal side effects and those with the greatest propensity for extrapyramidal side effects elicit the lowest incidence of sedation and hypotension. For young students and professional people who must be alert in their work, drugs such as the butyrophenones are first prescribed because they are least likely to impair alert behavior. If a patient is overly sensitive to such side effects, then the physician may switch to other drugs that are less likely to cause extrapyramidal effects.

Mechanism of action

In efforts to identify the mechanism whereby drugs exert their therapeutic action, pharmacologists frequently administer the drugs to animals and measure various physiological and biochemical parameters. In early studies with phenothiazines many different effects were observed. Phenothiazines are highly reactive chemical substances and influence all sorts of biochemical processes including protein synthesis, mitochondrial respiration, carbohydrate disposition, and the metabolism of all known neurotransmitters. The major task of researchers has been to distinguish which of these effects are responsible for the therapeutic actions of the drug and which are secondary or unrelated. The availability of an extensive number of clinically employed neuroleptics has facilitated such an evaluation. Scientists have compared the relative potencies of neuroleptics in producing various biochemical actions with their relative clinical potencies. Examined in this way, most of the reported effects of neuroleptics fail to correlate with clinical activity.

Research on neurotransmitter receptors has provided direct evidence for what is generally regarded as the therapeutic action of neuroleptics. In 1963 Arvid Carlsson observed changes in the metabolism of the neurotransmitter dopamine elicited with greatest potency by haloperidol, with lesser potency by chlorpromazine, and with no effect by the phenothiazine antihistamine promethazine, which lacks antipsychotic effects. Thus these initial observations reflected a correlation between therapeutic potency and biochemical effect. The pattern of change in dopamine metabolism suggested that neuroleptics accelerate the turnover of dopamine. Carlsson speculated that this augmented turnover reflected more rapid firing of dopamine neurons. He suggested that neuroleptics block postsynaptic dopamine receptors and that, by some feedback mechanism, the dopamine neurons fire more rapidly in an effort to overcome the receptor blockade.

When it became possible to measure dopamine receptors by monitoring the binding of radioactive neuroleptics to brain membranes, it was possible in our own laboratory as well as that of Philip Seeman in Toronto to show directly that neuroleptics block dopamine receptors and that the relative potencies of an extensive series of neuroleptics in blocking receptors correlates closely with their clinical potencies. This correlation holds for the D_2 subtype of dopamine receptors but not for D_1 receptors. D_1 dopamine receptors are linked to a stimulation of adenylate cyclase and were shown by Paul Greengard at Yale University to be blocked by phenothiazines potently but not by butyrophenone neuroleptics. Many neuroleptic drugs are potent blockers of other neurotransmitter receptors such as histamine H_1, $alpha_1$ adrenergic, muscarinic cholinergic, and serotonin S_2 receptors. However, effects on the other receptors do not correlate with therapeutic potency. Accordingly, it is generally accepted that D_2 dopamine receptor blockade accounts for the therapeutic actions of neuroleptic drugs.

Dopamine receptor blockade can also explain side effects of neuroleptics. Idiopathic Parkinson's disease is caused by a degeneration of dopamine neurons with cell bodies in the substantia nigra and terminals in the corpus striatum. By blocking dopamine receptors in the corpus striatum, neuroleptic drugs produce a functional dopamine deficiency, accounting for their parkinsonian side effects. Chronic blockade of dopamine receptors provokes a synthesis of new dopamine receptors by neurons receiving dopamine neuronal input. These new receptors cause a supersensitivity to dopamine. It is thought that this increase in dopamine receptor number can explain the symptoms of tardive dyskinesia.

Actions of neuroleptics at other neurotransmitter receptors are relevant to side effects of the drugs. The sedation and hypotension produced by neuroleptics is closely correlated with the extent of blockade of alpha-adrenergic receptors at therapeutic doses. Blockade of muscarinic acetylcholine receptors diminishes the propensity of neuroleptics to elicit acute pyramidal side effects. Thus, neuroleptics which are most potent in blocking muscarinic cholinergic receptors, such as thioridazine, produce the lowest incidence of acute extrapyramidal side effects. This mechanism fits with the well-established clinical evidence that anticholinergic drugs relieve extrapyramidal side effects of neuroleptic drugs. Moreover, anticholinergic drugs have been used with considerable success in the treatment of idiopathic Parkinson's disease.

Further reading

Creese I, Hamblin MW, Leff SE, Sibley DR (1983): CNS dopamine receptors. In: *Handbook of Psychopharmacology,* vol 17, Iversen, LL, Iversen SD, Snyder SH, eds. New York: Plenum Press, pp. 81–138

Lee T, Seeman P, Tourtelotte WW, Farley IJ, Hornykeiwicz O (1978): Binding of ^{3}H-neuroleptics and ^{3}H-apomorphine in schizophrenic brains. *Nature* 274:897–900

Moore RY, Bloom FE (1978): Central catecholamine neuron systems: Anatomy and physiology of the dopamine systems. *Ann Rev Neurosci* 1:129–169

Snyder SH (1984): Drug and neurotransmitter receptors in the brain. *Science* 224:22–31

Neuropharmacology

Jack R. Cooper

Neuropharmacology is the study of drugs that affect the nervous system where drugs are defined not only as agents that are foreign to the organism but include endogenous substances such as L-dopa. In the central nervous system (CNS), neuropharmacology encompasses the psychotropic drugs which alter mood and behavior, general anesthetics, analgetics, anticonvulsants, hypnotics, analeptics, and narcotics. In the peripheral nervous system, neuropharmacology includes drugs that block axonal conduction, neuromuscular blocking agents, and an extraordinary range of drugs that affect the autonomic and enteric nervous systems. Since neuropharmacology is not a specific discipline with its own technology such as neurophysiology or neurochemistry, neuropharmacologists, depending on their training, represent virtually every field of biomedical science. As long as their primary concern is with neuroactive drugs, investigators can choose to be classified as neuropharmacologists regardless of whether they are endocrinologists, immunologists, physiologists, or chemists.

Because of our neuroscientific naiveté at the molecular, cellular, and especially behavioral levels, most of the currently used neuropharmacologic agents have been discovered serendipitously: it is rare to encounter a therapeutic agent that has been the result of a rational drug search based on a fundamental discovery. An example of this rarity is the neuromuscular blocking drug, *d*-tubocurarine, discovered by Claude Bernard in his investigation of the mechanism of action of the poison used by Amazonians to coat their arrows. However, the increasing sophistication of neurophysiological, neurochemical, and neuroanatomical techniques has produced a quantum leap in our understanding of the nervous system that may make rational programs for drug development commonplace. To explain this statement, a little background information may be helpful. With the notable exception of local anesthetics which block axonal conduction, virtually all neuropharmacologic agents act on synapses to affect transmission. Historically, based on the discovery that eserine produced its acetylcholine-like effects because it inhibited acetylcholinesterase, which hydrolyzes acetylcholine, it was assumed that most drugs act by inhibiting an enzyme. It is now known, however, that this is a fairly uncommon event and that almost all drugs that affect the nervous system owe their activity to their interaction with a receptor. These receptors are found both presynaptically, where they alter the amount of neurotransmitter or neuromodulator released, and postsynaptically, where they alter the ensuing signal. Postsynaptic receptors can be affected by drug treatment so that the affinity of their endogenous ligand can be either reduced or increased and their density at the neuronal surface can be either decreased (down-regulated) or increased (up-regulated).

This receptor concept of drug action has revolutionized the production of new neuroactive drugs. With the availability of ligands of high specific radioactivity and specificity for receptors, it is now possible to rapidly screen new compounds by determining their ability to displace the radioactive ligand from its binding site to a brain homogenate or synaptosomal membrane preparation. Further, with the identification of receptor subtypes (e.g., there are now four adrenergic receptors classified as α_1, α_2, β_1, and β_2), it is now possible to synthesize agents that are relatively specific in their affinity for these subtypes and therefore relatively specific in their therapeutic activity. In addition, the identification of a receptor for morphine led to the search for a naturally occurring opioid and to the discovery of a whole class of opioid peptides with a widespread distribution in nervous tissue. The isolation of endorphins and enkephalins is now encouraging pharmaceutical houses to develop stable analogs of these endogenous antinociceptive peptides. Another major advance in drug synthesis that is attributable to the receptor concept is just emerging. That is, with the isolation and purification of receptors and a knowledge of their three-dimensional structure, at least at the active site, it should be possible with current computer technology to design conformationally specific agonists or antagonists. This not only would prevent the side effects of current drugs, but also would obviate the kind of molecular roulette that medicinal chemists now employ in rearranging structures of prototypical drugs. Finally, in the design of new drugs, one cannot fail to note the future impact on neuropharmacology of molecular neurobiology. It may be possible to combine monoclonal antibodies that recognize specific neural elements with drugs to produce a drug delivery system that will ensure a precise distribution of the drug to its appropriate site. In addition, recombinant DNA technology, with its ability to isolate, identify, and modify genetic material, may be the neuropharmacology of the 21st century. For example, Parkinson's disease is apparently caused by the destruction of nigrostriatal dopaminergic neurons, and is currently treated with L-dopa, the precursor of dopamine. Conceivably, gene replacement therapy may in the future substitute for L-dopa by producing a whole new set of neurons.

In general, drugs that affect the nervous system have traditionally been assumed to exert their activity by interacting with one of the neurotransmitters, i.e., acetylcholine, noradrenaline, dopamine, serotonin, gamma-aminobutyric acid, glycine, glutamate, or aspartate. It was thought that if we knew the circuitry involving these transmitters, we would understand brain function and, consequently, drug action. However, with the recent recognition of pre- and postsynaptic modulation of synaptic transmission, coupled with the discovery of close to three dozen neuroactive peptides in the brain, not to mention a number of small molecules such as adenosine, all of which primarily appear to modulate activity, our ignorance of brain circuitry has deepened. Since modulation confers a fine tuning of the nervous system, a requisite for elaborating behavioral changes, it is conceivable that the mechanism of action of the psychotropic drugs is referable to an effect on these modulators, or their receptors, particularly the peptides, and not directly on the neurotransmitters and their receptors. In one sense, this bewildering array of new endogenous neuroactive agents is depressing to brain theorists. To neuropharmacologists, however, it indicates the ability of the brain to regulate its activity using seemingly innumerable pathways and opens up many potential sites of drug intervention. Regardless of the difficulty in devising new agents for memory or aging, or control of neurological and mental dysfunction, neuropharmacologists will continue to use drugs as probes in an attempt to unravel the extraordinarily complex wiring of the brain.

Further reading

Cooper JR, Bloom FE, Roth RH (1986): *The Biochemical Basis of Neuropharmacology*. New York: Oxford University Press

Iversen S, Iversen LL (1981): *Behavioral Pharmacology*. New York: Oxford University Press

Phencyclidine

Robert C. Petersen

Phencyclidine, or PCP, chemically one of the arylcyclohexylamines, was originally developed as a dissociative anesthetic in the late 1950s. It was abandoned (for humans, but not other primates) in the mid 1960s because patients so often showed marked confusion and irrational behavior mimicking schizophrenia during postoperative recovery. Although PCP is the best known of about 30 psychoactive analogs, others, including PCE, TCP, PHP, PCC, and the shorter acting anesthetic, ketamine, are chemically related and have similar psychophysiological effects.

Unfortunately, the ease with which PCP could be synthesized coupled with the emergence of a psychedelic drug fad resulted in its widespread abuse beginning in the 1960s. Initially PCP developed a poor street reputation because overdoses were common when it was taken orally. However, illicit users soon discovered that by smoking PCP they could reduce the likelihood of many serious side effects while preserving the sought-after mind alteration.

Biological and behavioral effects

Effects on the central nervous system are as follows: (1) Low doses lead to "drunkenness" with numbness of the extremities. (2) Moderate doses produce analgesia; still higher doses anesthesia. (3) Psychic states resembling sensory isolation are produced, although markedly distorted sensory impulses do reach the neocortex. (4) Cataleptiform motor responses occur. (5) Large doses can produce convulsions and, as a result of respiratory depression, death. (6) Marked interspecies differences occur: In primates, including humans, depressant effects predominate, whereas in other species stimulant effects are common. Sympathomimetic effects in humans include increases in heart rate and blood pressure. Catecholamines are potentiated through a cocaine-like action.

Phencyclidine abuse

The exact extent of PCP abuse is not known. Since the drug so frequently masquerades as other drugs and goes by such a bewildering array of street names ("angel dust" is only one of many), even users are sometimes unaware they have taken it. National drug abuse survey data suggests use peaked in the 1970s and is now declining. The frequency with which serious side effects accompany recreational use is also uncertain. There is, however, good reason to believe that in some areas PCP has been a leading cause of inpatient psychiatric admissions. Since the drug's effects so frequently mimic those of an acute schizophrenic episode, their true source often goes undetected. Whether the drug directly induces the psychotic episode or whether it precipitates a psychosis in those who are already vulnerable is uncertain. Recurrent schizophrenic episodes without additional drug use as much as a year later have been reported. Serious depression subsequent to use is also common.

Since two of the effects of PCP are markedly distorted—paranoid thinking and reduced pain perception—violent behavior that is not easily controlled sometimes results. Death from accidental injury is a frequent concomitant of use. Swimming is reported by users to be especially pleasurable while PCP intoxicated, but the sensory distortions that the drug induces make such activity extremely hazardous, and drowning accidents are common.

Since there is still no specific antagonist for PCP, treatment varies, depending on the dose, from simple reassurance in a calming environment to intensive life support measures to combat the respiratory depression, convulsions, and coma that can lead to death from larger doses. Although detailed treatment recommendations are beyond the scope of this review, caution must be exercised. Depression and possible suicide attempts after the acute crisis is passed are a continuing hazard.

Whether difficulties with memory and thinking following chronic use persist and for how long remains unclear. Since PCP users frequently use alcohol and other illicit drugs as well, separating the chronic effects of PCP from those of other drugs is difficult. The possibility that PCP abuse may also affect the body's immune response has recently been raised by laboratory studies of cell cultures. Since PCP readily crosses the placental barrier, it is a potential hazard to the fetus. Infants exposed to the drug in utero have been characterized by sudden outbursts of agitation, increased lability, and poor consolability, although they did not differ from a control sample in scores at three months on the Bayley Scales of Infant Development.

Improved PCP detection techniques using homogenous enzyme immunoassay procedures make emergency determinations of serum levels more rapid and convenient. Since phencyclidine is a principal cause of emergency psychiatric hospitalization, routine testing for its presence in recent admission cases is indicated. Studies of PCP's pharmacokinetics provide a convincing rationale for vigorous acidification of urine and diuresis in treating phencyclidine intoxication.

Despite two decades of abuse, important questions about this drug remain unanswered. Although the more dramatic adverse consequences of use have received publicity, little is known about their frequency of occurrence. While we know primates will self-administer PCP and there is evidence of specific opiate/phencyclidine binding sites in the human brain, the neurophysiological basis for PCP abuse is still not well understood. Phencyclidine was initially viewed as prototypical of a potentially wider range of easily synthesized mind-altering drugs with marked potential for abuse, but this feared consequence has not as yet occurred.

Further reading

Chasnoff IJ, Burns WJ, Hatcher RP, Burns KA (1983): Phencyclidine: Effects on the fetus and neonate. *Dev Pharmacol Ther* 6(6):404–408

Clouet DH (1986): *Phencyclidine: An Update*, NIDA Reasearch Monograph, DHHS Publ No (ADM) 86–1443, Washington, D.C.: US Government Printing Office

Khansari N, Whitten HD, Fudenberg HH (1984): Phencylidine-induced immunodepression. *Science* 225(4657):76–78

Petersen RC, Stillman RC, eds (1978): *Phencyclidine (PCP) Abuse; An Appraisal*, NIDA Research Monograph, DHEW Publ No (ADM) 78–728. Washington DC: US Government Printing Office

Walberg CB, McCarron MM, Schulze BN (1983): Quantitation of phencyclidine in serum by enzyme immunoassay: Results in 405 patients. *J Anal Tox* 7(2):106–110

Phenylalanine and Mental Retardation (PKU)

William L. Nyhan

Phenylketonuria (PKU) is the most important and most frequent disorder of aromatic amino acid metabolism. It results from a defect in phenylalanine hydroxylase, the enzyme that normally converts phenylalanine to tyrosine. The disorder was discovered some 50 years ago by Fölling who observed a deep green color on the addition of ferric chloride to the urine of two mentally retarded siblings and characterized the compound responsible as phenylpyruvic acid, the phenylketone for which the disorder has been named.

PKU is the model disorder among genetically determined diseases. Early treatment to prevent the accumulation of phenylalanine and its metabolic products effectively prevents the clinical manifestations of the disease. Screening programs in which the phenylalanine concentration of the blood of the infant is assessed permit early diagnosis and treatment of the entire population of most developed countries and a public health approach to the control of genetic disease. More recently the gene for phenylalanine hydroxylase has been cloned and restriction fragment length polymorphisms have been demonstrated that permit heterozygote detection and prenatal diagnosis.

Clinical manifestations

Classic PKU is characterized by severe mental retardation. The IQ is usually below 50. The earliest symptom may be vomiting. It may be of sufficient severity to lead to a misdiagnosis of pyloric stenosis. Infants may also manifest irritability and a mild eczematoid rash. In an older untreated patient there may be severe itching. An unusual odor, caused by the accumulation of phenylacetic acid, has been described as mousy, barny, musty, or wolflike. Patients are blonde, blue-eyed, and fair in over 90% of cases. However, the disease is seen in dark-skinned people, including blacks, where the untreated patient is usually less deeply pigmented than other members of the family.

Neurological manifestations may be seen in a third of untreated patients who are spastic, hypertonic, and have increased deep tendon reflexes, but in only about 5% are these manifestations severe; some are athetoid. Another third have very mild neurological signs, such as a unilateral Babinski response. A third are without neurological manifestations. Seizures occur in about a fourth of patients and about 80% have electroencephalograph (EEG) abnormalities. Hyperactivity and problem behavior are common, as are purposeless movements and stereotypes.

Somatic development is usually normal, but the patient may be short. Minor malformations have been reported in some patients.

Biochemical features

Phenylalanine hydroxylase is an enzyme whose activity is expressed only in liver. In classic PKU the activity of this enzyme is virtually undetectable. Furthermore, no cross-reacting material is observed when reacted with antibody that detects the normal human enzyme. In the presence of a block at this level phenylalanine accumulates in body fluids. Concentrations in plasma virtually always exceed 20 mg/dl.

When phenylalanine accumulates it is converted to phenylpyruvic acid, phenyllactic acid, and phenylacetic acid. This compound is conjugated to form phenylacetylglutamine. Phenylalanine is also converted to *o*-hydroxyphenylacetic acid.

The diagnosis of PKU is made by routine screening of all infants in the first weeks of life. In the United States this is generally carried out on discharge of the newborn from hospital. It is desirable that testing be done after at least 24 hours of feedings containing protein. As many as 8% of patients may be missed because the screen is done before levels of phenylalanine have risen. A physician can be surer that his patient is not one of the 8% by retesting at 2–4 weeks.

The test is done on spots of dried blood on filter paper. A positive test maybe repeated or followed up directly with a quantitative determination of the concentrations of phenylalanine and tyrosine in plasma. Patients with PKU have markedly elevated concentrations of phenylalanine and usually reduced levels of tyrosine. In transient tyrosinemia of the newborn, especially common in premature infants, phenylalanine levels are high enough to trigger the screen but concentrations of tyrosine are always higher.

In order to distinguish the patient with classic PKU from some of the milder hyperphenylalaninemias that may not need treatment, it is useful to hospitalize the patient so that the intake of phenylalanine can be closely monitored and fresh urine specimens obtained. A standard challenge employed is 3 days of a 1:2 dilution of evaporated milk formula. In classic PKU the concentration of phenylalanine is over 20 mg/dl and usually over 30 mg/dl, and that of tyrosine is under 5 mg/dl. Patients with classic PKU will excrete phenylpyruvic acid and *o*-hydroxyphenylacetic acid. Therefore, the urine should be tested with $FeCl_3$ or chromatographed for phenolic acids. Rarely an immature infant may fail to excrete these metabolites. Thus any infant with a plasma concentration of phenylalanine over 20 mg/dl should be treated with a diet restricted in phenylalanine.

In the U.S. collaborative study, challenge with phenylalanine was carried out at 90–120 days and again after 1 year of age. The standard 3-day challenge of 24 ounces daily of 1:1 mixture of evaporated milk provides 180 mg/kg/day of phenylalanine for a 3–6-month-old infant. The dose is adjusted to 180 mg/kg/day for the 1 year old. In classic PKU the response to the challenge is a sharper rise in the serum concentration of phenylalanine to 30–40 mg/dl in 48 hours, at which level the challenge is terminated. This type of dosage should not be used for a diagnostic challenge in an older child where requirements are much lower than that of the infant. Symptomatic hypoglycemia has been reported with the use of this much phenylalanine in a 15-year-old patient.

Genetics

PKU is an autosomal recessive disease. Its frequency has become clear through genetic screening programs. In the United States, it approximates 1 : 15,000.

Heterozygote testing has been addressed by phenylalanine loading tests, but while accurate when positive, this approach is not reliable in excluding heterozygosity. In fact, assay of the activity of phenylalanine hydroxylase in biopsied liver may be confusing, in this case because of the inability to distinguish the heterozygote from the homozygote. The cloning of the gene and the assessment of restriction fragment length polymorphism has permitted the first reliable approach to the determination of the heterozygosity. It is not available in every family, but a majority of those tested are sufficiently polymorphic to be informative. In these families the method is also useful for prenatal diagnosis, and the prenatal diagnosis of an affected fetus has been carried out by Woo. Furthermore, this method should permit analysis of the samples obtained by chronic villus biopsy at 8–10 weeks of gestation, considerably earlier than is possible with aminocentesis where at least 2 weeks would be required to grow the cells following the procedure at 15 weeks.

Treatment

PKU is treated by providing a diet sufficiently low in phenylalanine that the concentrations of phenylalanine in the blood are maintained between 3 and 15 mg/dl. Infants detected in the neonatal screening programs and treated in this manner should develop normally and their IQs should be in the normal range. Preparations are available, such as Lofenolac (Mead-Johnson), which make infant feeding relatively simple, and lists of the phenylalanine contents of foods and of low protein foods are helpful in management. Treatment must be monitored by the regular determination of the blood concentration of phenylalanine. In this way one can avoid problems of poor compliance, on the one hand, and overtreatment with signs of phenylalanine deficiency, on the other.

The time at which dietary therapy may be discontinued is unclear. It appears that its stringency can be relaxed sometime between 6 and 15 years, but recent information indicates that some control should be continued through these years. Pregnancy in a female with PKU conveys a great risk of severe mental retardation in the offspring. Treatment of the mother during pregnancy is of only limited effectiveness. Treatment begun prior to conception may be more promising, but information is so far limited.

Other disorders of phenylalanine metabolism

The advent of the screening programs has brought to light some additional disorders that were previously not recognized. Among these are the so-called phenylalaninemias that appear to be defects in phenylalanine hydroxylase less severe than that of classic PKU. Because the enzyme is capable of partial activity, levels of phenylalanine are substantially lower than in classic PKU. In general these patients have serum concentrations of phenylalanine less than 20 mg/dl when the infant is receiving 180 mg/kg of phenylalanine. In contrast, the infant with classic PKU is unable to tolerate more than 75 mg/kg of phenylalanine. It is generally agreed that if patients with hyperphenylalaninemia have levels above 20 mg/dl they should be treated with phenylalanine restriction and if they do not it should be unnecessary to treat them.

Abnormalities in pteridine cofactor metabolism

Phenylalanine hydroxylase requires a tetrahydrobiopterin cofactor for the hydroxylation of the amino acid. In the course of the reaction, an oxidized quinonoid dihydrobiopterin is formed. This must be reduced to the tetrahydro form before it can be active again as a cofactor. The reduction is catalyzed by dihydropteridine reductase. The biopterin cofactor is synthesized from guanosine triphosphate through a series of steps in which neopterin and sepiapterin are intermediates.

Thus it is not surprising that genetically determined defects have turned up in the processes of cofactor metabolism that lead to hyperphenylalaninemia. Two general types of disorder have been described. In one there is defective activity of dihydropteridine reductase, and in the other the defect is in biopterin cofactor synthesis.

The clinical manifestations appear to be the same. Most patients have been recognized because of the presence of progressive cerebral disease despite excellent control of the plasma levels of phenylalanine. Tetrahydrobiopterin is also the cofactor for the hydroxylation of the tryptophan and tyrosine. Therefore, in these patients there should also be abnormal synthesis of serotonin, dihydroxyphenylalanine (DOPA), and norepinephrine, all neurotransmitters, and this could have relevance to the cerebral abnormalities. Involved patients are markedly hypotonic and spastic. They display distonic posturing. Mental retardation is profound, and drooling is a prominent feature. Seizures, including myoclonic seizures, are seen, and the EEG may be abnormal.

The diagnosis of these disorders can be made in a number of ways. The simplest is the administration of tetrahydrobiopterin. A dose of 2 mg/kg leads to a prompt decrease to normal in the concentration of phenylalanine in both patients with reductase defects and with synthesis defects. This is so simple that where the compound is available it should be given to every infant with a tentative diagnosis of PKU. The patient with PKU will have no change in the level of phenylalanine following administration of the cofactor. It is important that the patient be on a diet containing normal amounts of phenylalanine, not the therapeutic diet employed for patients with PKU. A few patients with reductase defects have been missed using the BH_4 loading test, but Danks has pointed out that his recommendation was for parenteral administration of BH_4 and the failures have been with the oral use of the compound. In those that respond, further workup consists in the assay of dihydropteridine reductase in cultured fibroblasts, lymphoblasts, or freshly isolated lymphocytes. Those who do not have reductase defects can be studied for biologically active pteridines using a bioassay with *Crithidia fasciculata*, high-voltage electrophoresis, or chromatography for the pteridines of the urine. Others have approached the detection of these defects in hyperphenylalaninemic patients by routinely examining the blood for BH_4 and studying the urine for pteridines using high-performance liquid chromatography (HPLC). This can be done with direct electrochemical detection in which each of the compounds is detected. More often the analysis is carried out of the oxidized pterins and various ratios of biopterin and neopterin expressed. Patients with synthesis defects fall into two classes. In those with a defect in the initial guanosine triphosphate (GTP) cyclohydrolase step all pterins are low in blood and urine. In patients with the more common defect in the area of dihydrobiopterin synthetase, the concentrations of BH_4 are very low and those of the neopterin high. The ratio of neopterin to biopterin is very high. In patients with reductase defects there appears to be a lack of feedback inhibition and so there is massive overproduction of urinary pterins. The level of BH_4 is always low, but high or normal levels of the other pterins have been observed. It may be that assay of the reductase enzyme is the only reliable way of detecting all patients with this defect. This can now be done on dried spots of blood on filter paper. Another possibility for the detection of synthesis defects is the measurement of tetrahydrobiopterin in plasma. It is high in PKU and increases in response to a phenylalanine load in normal individuals or those with PKU or with reductase defects but does not change in those with synthesis defects.

The treatment of these patients should be with tetrahydrobiopterin. In addition it has been recommended that these patients be treated with the serotonin and dopamine precursors, 5-hydroxytryptophan and DOPA. Some success has been reported with treatment, before the availability of the cofactor, with phenylalanine restriction and the administration of these biogenic amine precursors.

Further reading

Fölling, A (1934): Uber ausscheidung von Phenylbrenztraubensaure in den Harn als Stoffwechselanomalie in Verbindung mit Imbezillitat. *Hoppe-Seyler's Z Physiol Chem* 227:169

Kaufman S, Berlow S, Sumer GK, et al (1978): Hyperphenylalaninemia due to a deficiency of biopterin. *New Engl J Med* 299:673

Kaufman S, Holtzman NA, Milstien, S, Butler, IJ, Krumholz, A (1975): Phenylketonuria due to a deficiency of dihydropteridine reductase. *New Engl J Med* 293:785

Nyhan WL (1984): Phenylketonuria. In: *Abnormalities in Amino Acid Metabolism in Clinical Medicine*. Norwalk, Conn: Appleton-Century-Crofts, pp 129–148

Nyhan WL, Sakati NO (1976): Phenylketonuria (PKU). In: *Genetic and Malformation Syndromes in Clinical Medicine*. Chicago: Yearbook Medical Publishers, pp 6–9

Woo SLC, Lidsky A, Chandra T, et al (1983): Prenatal diagnosis and carrier detection of classical phenylketonuria by cloning and characterization of the human phenylalanine hydroxylase gene. *Clin Res* 31:479A

Premenstrual Syndrome

Stephanie J. Bird

Premenstrual syndrome (PMS) has become a widely recognized phenomenon by the medical, psychological, and scientific communities. Though more than 150 symptoms have been associated with it, those most often described are mood swings, breast tenderness, bloating, migraine headaches, irritability, and fatigue. Other symptoms commonly identified are tension, depression, anxiety, food cravings, abdominal discomfort, acne, edema, swelling of the extremities, and dizziness. Depending upon the symptoms and their severity, premenstrual changes may be simply an annoyance, or, if sufficiently severe or prolonged, they can be incapacitating.

PMS is not easily defined. It is clear that it is distinct from dysmenorrhea, the painful cramping associated with menstruation, and the result of elevated levels of prostaglandins. There is controversy regarding the number and combinations of symptoms, their time course, and their severity, as well as their etiology. Some researchers believe that there may be as many as four or five different types of PMS. Abraham has identified a major depressive syndrome including depression and insomnia, a water retention syndrome marked by edema and bloating, an anxiety syndrome, and an increased appetite syndrome. Endicott, Halbreich, and colleagues have also observed the major depressive and water retention syndromes, but describe a general discomfort syndrome, an impulsive syndrome, and an impaired social functioning syndrome rather than the anxiety and increased appetite syndromes. General discomfort syndrome includes headaches, back, joint, or muscle pain, and abdominal discomfort. Impulsive syndrome includes any two of the following characteristics: violent conduct, lack of self-control, impulsive behavior, and outbursts of irritability. Impaired social functioning syndrome is a complex of any three of eight symptoms including an increased tendency to "nag," mood changes, social withdrawal, and decreased activity. It should be noted that not all clinicians specializing in PMS find the symptom clusters described, but instead find that their patients report symptoms which cut across the groups, most frequently mood swings, migraines, breast tenderness, and bloating.

In theory, PMS is by definition premenstrual, i.e., symptoms recur cyclically prior to the onset of menses and cease with the beginning of menses. However, several variations in the pattern of symptom occurrence have been observed. First, symptoms may continue into the menses rather than ceasing with the onset of menstrual flow, and thus are more accurately para- or perimenstrual. Second, although symptoms may occur only pre- or paramenstrually, for some women symptoms occur at ovulation and either (1) disappear and recur premenstrually, (2) continue, gradually increasing in severity, until menstruation, or (3) continue at maximum severity throughout the luteal (postovulatory) phase. For a diagnosis of PMS, clinical consensus requires that prospective monitoring demonstrate the marked diminution, if not disappearance, of symptoms during the postmenstruum and until ovulation.

Premenstrual changes in physiological and psychological states are noted by the majority of women, although they are not always recognized as correlating with the luteal or premenstrual phase of the cycle. There is considerable variation in the severity of these changes and in the extent to which they are disabling. Some women experience severe, incapacitating, recurring symptoms. Many women notice detectable somatic and psychological changes which are not so severe that they are subjectively (or objectively) considered incapacitating or even sufficiently troubling to be thought of as a syndrome (i.e., indicative of an underlying disease process). Some women detect changes, such as increased energy, creativity, and sex drive, which they consider positive. A relatively small number of women note no changes at all. Thus, there is a continuum of symptomatology, and only those premenstrual changes sufficiently severe to be troubling or incapacitating can justifiably be identified as PMS.

Because of its link with the menstrual cycle, the majority of current hypotheses regarding the pathogenesis of PMS postulate some type of sex hormone-associated disorder. This seems feasible since steroid receptors have been localized to neurons in various brain regions and the influence of steroid hormones on the sensitivity of those neurons to other endogenous neuroactive substances has been demonstrated, i.e., part of the effect of these hormones is to modulate the effect, and thus action, of other neuroactive substances. In addition to the direct action of steroid hormones on neuronal function, however, there is increasing recognition of the role of other compounds associated with the hypothalamic-pituitary-reproductive organ axis, such as the gonadotrophins and releasing hormones, in central nervous system function. Thus, it is likely that sex hormones and other endogenous compounds associated with the reproductive cycle are involved in aspects of brain function not considered components of reproductive function.

Until recently a widely held view had been that menstrual disorders are psychological in origin, arising from personality maladjustment or neuroticism and lack of acceptance of the feminine role. In a critical review, Gannon has found that studies which putatively support that theory lack appropriate data upon which to base such conclusions. The precise nature and etiology of PMS are unclear. However a number of findings indicate it has physiological rather than psychological roots. There is common experience of premenstrual changes among women of widely divergent educational and cultural backgrounds, a continuum of symptom severity across individuals, and the observation of premenstrual changes in behavior in primates. Furthermore, the premenstrual, rather than menstrual, occurrence of PMS coupled with the fact that few women actually have 28-day cycles or even cycles of regular duration, and therefore rarely know precisely in which phase of their cycle they are (other than the menstrual phase)—make a purely psychological aetiology difficult to explain. However, discussion of somatogenic as opposed to psychogenic origins draws a somewhat artificial dichotomy: while hormones influence higher brain function including behavior, behavior affects hormone levels and activity. In addition, societal pressures have a direct influence on brain function through their effect on how the individual perceives her environment, the presence of threats and danger, her own ability to affect the world, and so on. Additional research is under way to clarify the effect of stress on PMS, the possible relationship of PMS to psychiatric disorders, and the influence of sociocultural factors. PMS, like other complex behaviors, is the result of dynamic interactions of various social and psychological, as well as neurobiological factors.

Further reading

Abplanalp JM, Haskett RF, Rose RM (1980): The premenstrual syndrome. *Adv Psychoneuroendocrinol* 3:327–347

Abraham GE (1980): Premenstrual tension. *Curr Prob Obstet Gyn* 3(12):1–39

Endicott J, Halbreich U, Schacht S, Nee J (1981): Premenstrual changes and affective disorders. *Psychosom Med* 43:519–529

Gannon L (1981): Evidence for a psychological etiology of menstrual disorders: a critical review. *Psychol Rep* 48:287–294

Reid RL (1985): Premenstrual syndrome. *Curr Prob Obstet Gyn and Infertil.* 8(2):1–57

Reid RL, Yen SSC (1981): Premenstrual syndrome. *Am J Obstet Gyn* 139:85–104

Psychiatry, Biological

Herbert Pardes

Biological psychiatry covers an extensive and rapidly changing area. Limited in the 1950s and early 1960s largely to psychopharmacology, some attempts at diagnosis, and early broad-brush attempts at explicating etiology (most notably through genetics), the advance of biological psychiatry grew steadily in the succeeding fifteen years, and today it is flush with excitement. What follows are some major trends and investigative lines.

Psychopharmacology

Prior to the use of drugs, psychiatric treatment was limited to convulsive therapies, some psychotherapies, removal of the patient from the ostensibly traumatizing environment, and various rest and rehabilitation strategies. The introduction of phenothiazines for psychotic symptoms allowed the return of state hospital patients to the community. Subsequent development of monoamine oxidase (MAO) inhibitors, benzodiazepines, tricyclics, and lithium enlarged the therapeutic armamentarium while bringing many side effects.

Benzodiazepines have proved useful for relief of anxiety and also as a substitute for sedative and hypnotic drugs. Addiction and withdrawal syndromes complicate their use. Current approaches (Shader and Greenblatt) stress nonpharmacological methods as treatment of choice in situational and nonincapacitating anxiety. When anxiety is severe and prolonged, adjunctive treatment is appropriate with antidepressants for phobic and panic states, and anxiolytics (e.g., benzodiazepines) for generalized anxiety disorder and mixed anxiety and depression. While benzodiazepines are currently the drug of choice among anxiolytics, newer nonbenzodiazepine compounds may provide increased safety.

The tricyclics enjoy preeminence for the treatment of depression, though lithium (which is effective in 70% of cases of manic-depressive illness) is available along with various proven psychotherapies. While lithium is especially helpful in preventing recurrences of mania, tricyclics have recently been shown far more effective in preventing recurrences of depression in unipolar depressives and in treating the acute unipolar depression itself. Adding lithium to a tricyclic (e.g., imipramine) does not increase effectiveness. Lithium is as effective as imipramine in preventing recurrent depressions in a bipolar illness, but adding imipramine to lithium does not increase effectiveness.

Choice of treatment for depression is especially important in the elderly for whom depression is the most common emotional disorder. The many second-generation drugs now available makes picking correct treatment more complicated.

MAO inhibitors have enjoyed a revived use after earlier disfavor due to side effects. They are now being invoked increasingly for affective disorders, anxiety states, and phobias.

For schizophrenics pharmacologic agents (e.g., phenothiazines) can alleviate the most disruptive symptoms (hallucinations and delusions), but for interpersonal difficulties, psychosocial approaches focusing on the family appear necessary along with attention to the patient's psychosocial needs (work, income, housing friends, etc.). This use of multiple interventions has its parallel in the use of combined pharmacotherapeutic and psychotherapeutic intervention in anxiety and depressive disorders.

Tardive dyskinesia is an omnipresent liability in the use of phenothiazines. A syndrome characterized by involuntary movement of face and extremities, it is estimated to affect 14% of patients who have four years of cumulative neuroleptic exposure. Theories that it is caused by dopamine hypersensitivity remain to be proved, and various treatment strategies (intermittent treatment, low doses) are being explored.

Electroconvulsive therapy is used for some patients with affective disorders and for some catatonic patients. Carbamazepine, a drug previously used as an anticonvulsant, has been effective in some manic-depressive patients who are unresponsive to lithium.

Psychopharmacology may benefit from such new technology as positron emission tomography (PET) which may determine in advance the relative action of new drugs on major target receptors versus their unintended influence on other receptors. The explanation, mitigation, and prevention of tardive dyskinesia and the sorting out of differential values of various drugs for specific conditions (even antidepressants now have been found to have substantial antianxiety effects) are major themes in psychopharmacology research.

Diagnosis and epidemiology

The third edition of the *Diagnostic and Statistical Manual on Mental Disorders* (DSM-III) of the American Psychiatric Association, the recently introduced system for psychiatric diagnosis, reduced the stress on clinical inference in favor of observable phenomena as a basis for diagnosis. While challenged as discarding the value of inference and implying closure on diagnosis prematurely, DSM-III has increased the reliability of psychiatric diagnosis. It exemplifies a move in psychiatry to greater refinement and differentiation generally. Using refined interview techniques and scales based on symptoms and disease syndromes rather than on general psychological function and well-being, psychiatric epidemiology has been developing more solidly based data on the incidence and prevalence of the mental disorders estimated to plague 33 to 38 million Americans.

Subtyping of affective disorders has been followed by subtyping of anxiety disorders. Some evidence suggests that panic anxiety is qualitatively separable from other anxiety disorders. Vulnerable subjects may have attacks precipitated with laboratory infusion of lactic acid. Furthermore, antidepressants may selectively suppress panic attacks.

Subtyping schizophrenia has been generally acknowledged

as a commendable research strategy. Still, no biological marker has been fully accepted in psychopathology. Perhaps shifting the biological marker from the dependent variable to an independent criterion would allow new diagnostic entities created around a marker rather than around symptoms.

Laboratory tests, hitherto unused in psychiatric diagnosis, appear increasingly likely to prove of value. The dexamethasone suppression test (DST), a test based on the existence in many depressed patients of an abnormal cortisol response to the synthetic steroid hormone dexamethasone, while of little value in general medical settings, may have some value as a confirmatory test in endogenous or major depressive disorders. Sensitivity and specificity vary greatly, however, in different clinical settings. Though currently limited in value DST, thyrotropin-releasing factor (TRF), and possible urinary tests for selecting appropriate treatment in depression all point to increasing laboratory participation in clinical psychiatry.

Psychiatric disorders

Nearly one in four women and one in ten men will suffer depression some time in their lives. This is a treatable disorder with estimates of 80–90% of depressions responding to intervention. The most tragic complication is suicide, which befalls some 15% of people suffering from depression. Suicide has risen dramatically in adolescents between 15 and 24, young urban blacks, and the elderly. The course of depression has been found to be more pernicious than earlier thought with about 20% of people with depression continuing to be ill 2 years later and another 30–40% having a recurrent episodic course.

Attempts to understand the etiology of depressions have evoked a continuing debate as to the importance of endogenous versus environmental factors. It is suggested that some individuals have a genetic predisposition to develop affective disorders. Attempts to tie psychiatric disorders to neurotransmitter correlates is so far seen as generally simplistic.

The tie between the brain pathways that use norepinephrine as a transmitter and depression has been bolstered by experimental evidence. However, the evidence has suggested various mechanisms to account for the relationship. These range from decreased function to overactivity for norepinephrine-using neural pathways. The continuing accretion of findings substantiating such a relationship, yet appearing at times to be directly contradictory as to the mechanism, account for some of the pessimism and catalyze the examination of still other ways by which the norepinephrine system and perhaps other neurotransmitter systems may be linked with depression. There is hope newer mechanisms for determining genetic markers may catalyze the investigation of the etiology of affective disorders.

There is considerable evidence that children of parents with affective disorders are at significant risk for a variety of behavioral abnormalities out of proportion to other children in the population.

Schizophrenia, which affects some 1% of the population, is the core psychiatric disorder that probably attracts the most public attention. Deinstitutionalization policies, which offered some benefit in getting long-standing inhabitants of state hospitals back into the community, have also had the effect of pouring into the streets a large number of homeless people. Furthermore, the disease, which affects young adolescents and young adults particularly at the onset, is crippling in function and evokes public attention by virtue of its florid symptomatology.

Genetic predisposition to some schizophrenias seems sure. However, some familial and epidemiological evidence supports the idea of transmission from affected to nonaffected individuals. A viral transmission has been postulated in individuals who may otherwise be genetically predisposed. Other suggestions include that of an underlying autoimmune mechanism.

As mentioned earlier, subtyping of schizophrenia with an attempt to delineate smaller and more homogenous categories of disease that might be more readily susceptible of solution is an important current research perspective. Also treatment has assumed a multimodality character, given the failure of drugs alone to control symptomatology and restore function.

Alzheimer's disease, a neuropsychiatric disorder affecting some 6% of people over 65 and increasing to as high as 20% in the mid-80s is also the most common cause of mental deterioration. With a projected increase in the over-65 population to some 18% of the U.S. population from the current figure of 11%, the projection is for larger numbers of patients with Alzheimer's. This disease attacks many cognitive functions and perniciously nullifies the personality. As many as 15% of the cases represent a misdiagnosis of depression; this is important because depression is treatable.

Some patients with Alzheimer's disease may have a genetic basis. Other research has shown that acetylcholine-releasing neurons with cell bodies in the basal forebrain selectively degenerate in Alzheimer's disease. These cholinergic neurons innervate widely the cerebral cortex and related structures and apparently play a major role in cognitive functions, especially memory.

Attempted treatments to this point have been only effective in modifying some of the secondary symptoms such as the behavioral disturbances which are part of the disorder. For the basic cognitive deterioration, attempts to capitalize on the cholinergic theory with treatments that restore acetylcholine activity have to this point proved ineffective. An interesting similarity in the pathology found in older patients with Down's syndrome and those with Alzheimer's disease has been noted, but its significance is not understood.

These are a sampling of the many clinical psychiatric disorders. The principles of subtyping and increasing refinement and differentiation characterize the current evolution of clinical psychiatry. This refinement should be fostered further by the many dramatic developments in neuroscience such as imaging techniques, molecular genetics in its application to neurobiology and behavior, the enhanced capacity for detecting neurotransmitters, mapping and quantifying receptors, and the emerging research on the biology of higher functions. This latter focus on the interface of biology and behavior is of particular consequence for psychiatry since the disorders with which it deals are largely abnormalities of behavior. The vigor of the neuroscience developments with their implications for understanding behavior and fostering its favorable modification suggests that the next 10–20 years will show substantial advance in the differentiation of psychiatric concepts and clinical application.

Further reading

Buschsbaum MS, Haier RJ (1983): Psychopathology: Biological approaches. *Annu Rev Psychol* 34:410–430

Coyle JT, Price DL, Delong MR (1983): Alzheimer's disease: A disorder of critical cholinergic innervation. *Science* 219:1184–1190

Goldin ER, Gershon EJ (1983): Association and lineage studies of genetic marker loci in major psychiatric disorders. *Psychiatr Dev* 1:387–418

Kandel ER (1979): Psychotherapy and the single synapse: The impact on psychiatric thought on neurobiological research. *N Engl J Med* 301:1028–1037

Psychoanalysis

Charles Brenner

Psychoanalysis denotes a form of therapy that is, at the same time, a source of psychological data. The word also denotes the theories or generalizations derived from those data. As an investigative technique, the psychoanalytic method seems the best method currently available for the study of a certain aspect of the functioning of the human cerebral cortex, namely, the aspect which includes ideas, thoughts, plans, fantasies, emotions—in brief, whatever persons feel to be of prime importance to them in their mental lives.

Before the development of the psychoanalytic method, these data, usually called subjective data, were observable in a systematic way only by introspection. The findings of introspection were obviously unreliable, but the reason for their unreliability was not apparent until the psychoanalytic method was introduced. Data made available by the psychoanalytic method show that all persons are at pains to conceal or disguise from themselves many important motives for thought and behavior. Thus the data of introspection have been systematically falsified by the observer, whatever effort the observer may have made to report them honestly.

The psychoanalytic method is superficially simple. Two people meet daily for a convenient period of time, usually 50 minutes, in a setting designed to minimize extraneous stimuli: a quiet room with a couch on which the subject (analysand) can lie comfortably and a chair on which the observer (analyst) can sit out of range of the analysand's vision. The analysand speaks, the analyst listens. In addition, there is an agreement between the two that is of fundamental importance to the method itself. The analysand undertakes to tell without reserve whatever he is aware of, whatever is passing through his mind, whether it be thoughts, emotions, or physical sensations. The analyst agrees to restrict himself to listening, to trying to make sense of what he hears and otherwise observes (bodily movements, tears, laughter, tone of voice, etc.) and to communicate his conclusions to the analysand from time to time. It is essential to add that the analysand is a patient and the analyst a therapist. The one comes to the other for help with whatever psychogenic distress (symptoms) he is suffering from. The analyst offers the prospect of cure or improvement. The application of the psychoanalytic method is of necessity limited to a therapeutic situation for an obvious reason. It is only in a therapeutic situation that one person will speak to another without reserve for 50 minutes a day for an extended period of time.

It is apparent that the data made accessible by the application of the psychoanalytic method are the same in kind as the data of introspection. They are subjective, psychical data. The difference between the two methods is that in the one case the data are observed by the subject, in the other, by an independent observer (analyst). The difference seems, a priori, unimportant, but, in fact, it is crucial. An independent observer, provided he is experienced and unprejudiced, can learn far more about the aspects of mental functioning that are subjectively important than anyone can learn by introspection. During the century since Freud devised the psychoanalytic method and began to apply it, more has been learned about the mind than during all the centuries that preceded Freud's work. At present the psychoanalytic method is the best means we have for studying this aspect of the functioning of the human cerebral cortex.

Psychoanalysts assume that the mind is like every other aspect of mass/energy in space/time in that it is accessible to scientific study in the usual way. Although psychoanalytic data are subjective, psychoanalysts deal with them in ways which are no different in principle from the ways in which any scientific investigator deals with data. Psychoanalytic generalizations or theories are like any other scientific theories with respect to their relationship to data of observation.

Among the findings of psychoanalysis, the ones of most general importance have to do with psychic determinism and with unconscious mentation. Other important generalizations have to do with childhood sexuality, with the importance of pleasure seeking (libidinal) and aggressive motives in mental life, and with the importance of psychic conflict related to libidinal and aggressive wishes originating in childhood. Because it began as psychotherapy, psychoanalysis was, at the start, primarily concerned with the nature and origin of psychogenic symptoms. Thus its theories about sexuality and psychic conflict were, at first, discoveries about psychopathology. Their significance was soon extended, however, and by now the psychoanalytic theory of conflict relates to normal psychic phenomena as well as to pathological ones. The study of the psychic significance of dreams, of everyday slips and errors, and of jokes were of particular importance for this development.

The psychoanalytic method continues to be a specialized form of psychotherapy which requires a course of specialized training to master. The findings (psychological theories) of psychoanalysis, however, are of very general significance. A knowledge of them is indispensible to all who are seriously concerned with humankind and its works, i.e., to students of behavior, of the social sciences, of art, of literature, and of general psychology no less than to psychotherapists.

Further reading

Brenner C (1973): *An Elementary Textbook of Psychoanalysis*, 2nd ed. New York: International Universities Press

Brenner C (1982): *The Mind in Conflict*. New York: International Universities Press

Psychopharmacology

Susan D. Iversen

Psychopharmacology is concerned with the effects of drugs on animal and human behavior. It encompasses any chemical that influences behavior by a direct or indirect effect on the central nervous system. Drugs can be self-administered to alter normal behavior or clinically administered to control abnormal behavior; they can be administered acutely or chronically; and they can prove toxic in the long term. Psychopharmacology is a relatively new but rapidly developing branch of natural science that evolved from the discovery that organisms do not behave randomly but lawfully as defined by the structure of the environment. In much the same way that antihypertensive drugs modify the relationships of the cardiovascular system and control harmful high blood pressure, so psychoactive compounds result in orderly modifications of behavior. There are a number of outcomes to this discovery: the classification of drugs by their effects on normal animal or human behavior, rather than by physiological, biochemical, or pharmacologic criteria; the use of drugs as tools for probing the normal organization of psychological processes underlying behavior in animals and humans; and a rational use of drugs to control and restructure abnormal behavior.

Learning theorists have been defining the parameters of behavioral control throughout the 20th century, but B.F. Skinner introduced the technology for precisely establishing and measuring learned responses. The free operant method, derived directly from Skinner's work, depends on quantifying rates of performing a particular act, like lever pressing or key pecking. Rates of responding can be sensitive indices of motivation, emotion, arousal, and motor performance and of the effects of drugs on such mechanisms. Reinforcement and punishment or, in everyday terms, pleasant and unpleasant events determine the rate and pattern of conditioned responding. The probability of occurrence of a response increases if followed by a reinforcement, but if followed by punishment its occurrence decreases. Though an animal in a free operant situation is free to respond at any time, the schedule on which the reinforcing or punishing event is presented determines the subsequent rate and pattern of response. A fixed-interval schedule of reinforcement (reward at regular intervals of time) results in pauses followed by bursts of responding before the next reward. Ratio schedules (reward after a set number of responses), however, generate sustained high rates of response with very brief post-reinforcement pauses. Multiple schedules can be programmed to generate varying rates and patterns of response over time.

Psychoactive drugs from different classes modify schedule controlled behavior in characteristic ways, and if a battery of schedules involving rewarding and aversive stimuli is used a profile for a particular drug class can be defined. For example, stimulant drugs like amphetamine and minor tranquilizers, like chlordiazepoxide (CDP) or diazepam, change behavior maintained by reinforcement schedules in opposite ways; amphetamine enhances low response rates on interval schedules but does not release responses suppressed by the presentation of punishing stimuli, like electric shock. The benzodiazepine minor tranquilizers have the opposite profile. Accordingly, punishment procedures are essential for screening new anxiolytic drugs.

To cite another striking example, electric shock presented on schedules of punishment suppresses behavior, but in different circumstances animals learn a new response to terminate or avoid the same shock (negative reinforcement). Chlordiazepoxide (CDP) abolishes the suppression of response to a punishing stimulus but the antischizophrenic drug chlorpromazine does not. In contrast, the antipsychotic drug impairs escape and avoidance behavior, but CDP does not.

However, some drug effects are not readily reflected in changes in the response rate or pattern. Within this category are drugs which cause changes in sensory thresholds, discrimination, and memory processes. Here it is sufficient to present a stimulus and to determine the probability and accuracy of the response to that stimulus. Failure to detect the stimulus, discriminate it from others, or to retain the information results in loss of performance. The physical characteristics of the stimulus or the period of time for which it has to be retained can be varied to assess the limits of threshold, discriminability, and memory. Discrete trial procedures serve these purposes.

In these procedures, the animals are not free to respond at all times, but instead, the exposure to the situation in which the animal can respond is divided into discrete periods, called trials. Typically, one or more stimuli are presented in conjunction with each trial, and performance is measured in terms of the probability of emitting the correct response in relation to those stimuli. Such methods are widely used in human experimental psychology, and sophisticated methods of analyzing response probabilities are available.

Many psychoactive drugs influence the visceral nervous system. It is known that this sensory information reaches the brain and no doubt contributes to the accuracy with which animals and humans are able to distinguish different drug states. Unconditioned visceral stimuli come to modify subsequent behavior by the process of classical conditioning; after pairing of a neutral signal with an unconditioned stimulus which elicits a specific unconditioned response, the neutral stimulus comes to elicit that response. Attention is now being given to drugs as unconditioned stimuli, but equally important is the effect of drugs on established classically conditioned responses of the autonomic nervous system.

Finally, the study of such instinctive unlearned behaviors as feeding, drinking, mating, rearing young, arousal, aggression, and exploration, all implicated in fundamental homeostasis, provides a new impetus for psychopharmacology. Direct observation is the most appropriate for studying the effects of drugs on such mechanisms. In many neuropsychiatric conditions disturbances of motivation and emotion are more dramatic than those of cognition.

Twenty years ago very little was known about the effect

of psychoactive drugs on the biochemistry and the electrophysiology of the brain. The revolution that has taken place in neuropharmacology has shown how many psychoactive drugs modify the chemical and electrical properties of brain. In the early days of behavioral pharmacology it was neither necessary nor possible to implicate real brain mechanisms in the account of how drugs lawfully modified behavior. The challenge now is to complete the equation and understand how drugs influence the brain mechanisms underlying behavioral organization. A number of animal models have developed to predict psychoactive drug actions in humans. For example, models in which basal ganglia dopamine function is depressed, enhanced, or unbalanced by dopamine receptor antagonists, used to treat schizophrenia, provide a sensitive behavioral index for comparing the potency of novel dopaminergic drugs.

The revolution in neuropharmacology has also encouraged the view that since therapeutically useful drugs interact with neurotransmitters in brain, it is possible that neurochemical imbalance underlies the neuropsychiatric conditions in which they are beneficial. It seems worthwhile therefore to use such knowledge to develop animal models of the abnormal behaviors seen in illness. Psychoactive drugs may then be evaluated for their ability to restructure abnormal behavior patterns. Pharmacologic, neurological, or environmental variables may be manipulated to modify and disrupt behavioral baselines. For example, chronic amphetamine administration in a range of species, but especially in monkeys, induces a form of psychotic behavior likened to schizophrenia; electrical stimulation of the noradrenaline neurons of locus coeruleus in monkeys induces a behavioral profile including intense fear reactions and anxiety; and chronic unavoidable shock induces a form of behavioral despair, characterized by immobility and withdrawal which has been likened to human reactive depression.

Whether psychoactive drugs are used to enhance performance or control abnormal behavior, the response to the drug does not remain constant over time. Psychological tolerance is the most striking example of this, which results in larger and larger doses of the drug being required to achieve the same behavioral response. But equally important, the environment in which the drug is experienced becomes part of the tolerance syndrome. Physiological tolerance to the analgesic effect of morphine is readily demonstrated, but if morphine is administered repeatedly in a highly discriminable environment and analgesia testing is performed in an equally discriminable but different environment, the degree of tolerance seen on the pain threshold response is significantly reduced. It is believed that the change in the drug response becomes classically conditioned to the stimuli of the environment. These findings have important implications for the control of human drug addiction since it is claimed that withdrawal from the drug is unlikely to succeed if the stimuli conditioned to the drug state (e.g., syringe and needle paraphernalia, street corner drug scene) continue to influence the drug-seeking behavior of the addict.

Many other variables influence the response to a psychoactive drug; tolerance to the physiological effects of the drug, genetic makeup, psychological set at the time the drug is taken, behavioral history, and experience with other drugs. Although it is difficult to predict drug response, the technology of behavioral pharmacology provides the means to do so.

Further reading

Cooper JR, Bloom FE, Roth RH (1982): *The Biochemical Basis of Neuropharmacology*, 4th ed. Oxford: Oxford University Press

Heise GA, Milar KS (1984): Drugs and stimulus control. In: *Handbook of Psychopharmacology*, vol 18, Iversen LL, Iversen SD, Snyder SH, eds. New York: Plenum Press, pp 129–190

Iversen SD, Iversen LL (1981): *Behavioural Pharmacology*. New York: Oxford Press

Morse WH, McKearney JW, Kelleher RT (1977): Control of behavior by noxious stimuli. In: *Handbook of Psychopharmacology*, vol 7, Iversen LL, Iversen SD, Snyder SH, eds. New York: Plenum Press, pp 151–180

Wikler A (1965): Conditioning factors in opiate addiction and relapse. In: *Narcotics*, Wilmer DM, Kassenbaum GG, eds. New York: McGraw-Hill, pp 85–100

Psychosis

Gardner C. Quarton

A psychosis is a human behavioral state characterized by delusions, hallucinations, incoherent communication, loosening of association between ideas, markedpoverty of content of thought, markedly illogical thinking, or behavior that is grossly disorganized or catatonic. The term is unsatisfactory for scientific purposes.

A brief review of the history of the use of the term will demonstrate shifts in meaning that are still confusing today. In late Greek *psychosis* was used to connote animation or the principle of life. Greek and Roman physicians used terms based on other roots, e.g., *insania* and *furia,* to refer to mental illnesses we would label as psychotic. During the middle ages and even up to the beginning of the 19th century, individuals, today regarded as ill, were believed by Europeans to be possessed by the devil. With the increasing impact of rationalism and the scientific method, on the one hand, and humane concern for the unfortunate, on the other, the recognition developed gradually that the condition of these individuals need not be explained by supernatural phenomena. This induced the doctors of the time to look for causes of mental illness in correlations between the clinical manifestations of illness and pathology recognized in the tissues of the body at autopsy.

However, the effort to identify tissue pathology correlated with the behavioral disorders created a special problem not present for other illness. The abnormal thinking of the individual, who 100 years later would be called psychotic, seemed to involve another ontological category, the realm of the mind or spirit. In other words, the same factors in Western cultural and intellectual history that led to the distinction between mind and body, and to the distinction between materialism and idealism, created a need to distinguish a substrate that was pathological in ordinary bodily disease from that substrate that was pathological in mental disease. In the first half of the 19th century, the term *neurosis* referred to an abnormal state of the nervous system, i.e., of the body, and a correlative term, psychosis, referred to an abnormal state of the mind.

Four features of the use of the terms neurosis and psychosis in much of the 19th century must be kept in mind. First, the user needed to believe that the mind was, in some metaphysical sense, a different realm of being from the body. Second, the one who employed the word psychosis did so with the understanding that the psychoses exhausted the domain of psychological pathology. Third, this definition of psychosis is connotative rather than denotative, and it does not help in defining the lower threshold of unusual behavior that establishes abnormality. In other words, the normative problem is ignored and the need to create a clear criterion for the threshold is evaded. Last, the psychiatrist of the 19th century was quite comfortable inferring pathology in either the mind or nervous system when the evidence for that pathology was not particularly good.

Unfortunately this uniform pattern of usage was not to last. Late in the 19th century a new group of mental disorders began to be studied. These included pathology of sexual behavior such as the perversions, episodes of severe obsessive compulsive thinking and behavior, phobic behavior, episodes of anxiety and panic, and some disorders of memory, sensation, and motor function, strongly influenced by suggestion and hypnosis. This group came to be called the psychoneuroses in contrast to the other disorders of behavior known as psychoses. By the mid-20th century psychoneurosis was shortened to neurosis, and the neat nomenclature of the 19th century was destroyed, leaving in its place confusion over mind-body issues, the intent of the user with respect to the importance of nervous system changes etiologically, and with many of the connotations of the older usage pattern arising intermittently and irregularly.

This shift in usage required a set of criteria for distinguishing a psychotic state from a neurotic one. Three different types of criteria were used between 1950 and 1980, sometimes separately, sometimes combined, but rarely with clarity. Each of these can be found in medical dictionaries.

The first criterion is based on the assumption that a given episode of illness can be ranked on a single scale of seriousness. The psychotic states are serious; the neurotic, less so. This usage is unsatisfactory because it is difficult to identify a single dimension of seriousness along which all mental illness can be ranked, because it does not provide an easy and operational way of establishing the cutting line that divides the neuroses from the psychoses, and because some neurotic states seem at times to be more serious than some of the milder conditions still identified as psychoses.

The second criterion is based on the concept of the presence or absence of insight. The psychotic individual is said either not to know that he is sick, or if he understands that others believe him to be sick, he does not share this belief. The neurotic individual, in this usage of these terms, has some understanding that his condition is pathological; however, he still cannot control it, and his suffering and inconvenience are real. This criterion is also unsatisfactory for scientific purposes because the presence or absence of insight is both inferential and not easy to justify on the basis of evidence. It is difficult to establish any set of operations that provides an unambiguous cutting line that says how little insight you have to have to be called psychotic.

The third criterion is based on the presence or absence of specific signs or symptoms of the abnormal condition. The third edition of the *Diagnostic and Statistical Manual of Mental Disorders* of the American Psychiatric Association (1980) (DSM III), for instance, identifies as psychotic symptoms, the following: delusions, hallucinations, incoherence, loosening of associations, markedly illogical thinking, and behavior that is grossly disorganized or catatonic. This criterion has the advantage that it represents an effort to establish specific conditions that have to be met if an illness is to be called a psychosis. It has the disadvantage that it is a disjunctive definition. One individual may be declared psychotic because he has delusions but is otherwise apparently normal; another, because he shows disorganized behavior. This can lead to subsets of patients that may have quite different clinical pictures and prognoses, and where the etiological patterns may turn out to be based on different classes of determinants. Furthermore, each of these criteria is based on inference, and there is a normative element that may depend on the culture and educational level of the user. Furthermore, failure to detect the presence of one of these features does not always mean it is not present.

Behaviorism (particularly radical behaviorism), positivism, and analytic philosophy, in the first half of the 20th century, were strongly antimetaphysical, and therefore, those influenced by them attempted to solve the mind-body problem in a nonmetaphysical fashion. The separation of mind and body was treated as a category mistake. (It would be a category mistake to believe that a university and the buildings of that university were two separate entities rather than two linguistic ways of describing the same entity.) Under this interpretation, the creation of an entity called *mind* is a category mistake. Delusions, and the other defining features of psychosis, are considered patterns of function of the brain.

Sociologists have pointed out that whether a person in a given society is called psychotic or not is a function of the characteristics of the response to that person by the other people in the society. Erving Goffman has argued that a psychotic person fails to follow the complex rules that govern verbal and nonverbal interpersonal interactions, particularly between strangers. This may be because the individual does not know the rules, knows them but is, for some reason, incapable of following them, or simply does not follow them because he is engaged with an agenda of his own that is incompatible with his paying attention to these rules. The observer may not be able to discriminate among these, and that is why foreigners and handicapped persons are often thought, erroneously, to be suffering from a mental disease. A small group of sociologists and psychiatrists have taken the position that there is no such thing as mental illness; there is only the labeling of patterns of behavioral deviance. This extreme position ignores the brain events that many scientists believe are necessary to abnormal behavior. This argument is challenging, however, because it calls attention to the limits of behaviorism by making it clear that the patterns of behavior of the psychotic individual that seem odd may be based on the discrimination by the observer of very subtle peculiarities in interpersonal interaction rather than on the detection of well established signs of illness, and that this discrimination will depend on the cultural pattern of both the psychotic and the observer, on the context in which the odd behavior is emitted, and on the sensitivity of the observer to possible determinants of the deviations from the expected. This subtlety of discrimination may be incompatible with the reliability and operational explicitness of sorting behavior expected in modern science.

Explanation of psychosis

No general explanation of psychosis exists at the present time that identifies the conditions that are necessary or sufficient for the manifestation of all patterns of behavior we usually call psychotic.

It is convenient to divide hypothetical determining events into those which are, in some sense, input into the organism, and those which are the intervening patterns of brain activity that are the basis for the normal or psychotic behavior. Among the important input events, we may consider: (1) selection of genes at conception, (2) factors in the fetal environment influencing brain development, (3) factors in the postfetal brain environment that alter brain structure or function, e.g., drugs, infection, nutrition, tumors, trauma to brain, etc., and (4) factors in the postfetal environment that act through sensing and coding systems of brain, e.g., sensory deprivation, meaning deprivation, deficient or atypical learning, or stress—all conditions that can alter brain programming.

In manic-depressive and schizophrenic psychoses there is evidence for a role of genetic factors. The presence of a specific genotype, however, is insufficient to completely determine the psychotic behavior because individuals who have inherited the required genotype do not always manifest psychosis. It is not known whether all schizophrenic-like individuals have an abnormal gene product playing an etiological role. The genes play a predisposing or determining role in some but not all the other psychoses.

Factors in the fetal environment that influence brain development, if they affect behavior, are most likely to produce mental retardation syndromes. Sometimes this retardation is accompanied by episodes of psychosis. In the absence of retardation, disorders of development probably contribute to psychosis later in life, but clear-cut evidence for this is not yet well established.

Environmental influences after birth, affecting brain directly, are well known to produce psychosis, even if there is no evidence for a psychotic genotype. Brain tumors in the frontal and temporal lobes are sometimes associated with psychotic behavior. Other forms of gross brain pathology, particularly in the temporal lobe and when accompanied by seizures, can produce abnormal behavior often misdiagnosed as schizophrenia. Neurosyphilis was a common cause of psychotic behavior characterized by grandiose delusions until about 1940 when treatment produced a dramatic reduction in prevalence of neurosyphilis. Meningitis and encephalitis due to a variety of organisms occasionally produce psychotic behavior. Nutritional, metabolic, and endocrine disorders sometimes become important determinants of odd behavior. A large number of drugs, acting on the brain in different ways, can contribute to the development of psychosis. One of the most interesting of these is amphetamine, which can produce psychotic signs in some rare individuals with doses as low as 20 mg but which will do so in almost any individual if the dose is sufficiently high. Some drugs that produce psychosis act on the catecholamine transmitter systems, others are anticholinergic, still others are sedative. Many drugs that can produce a psychosis are not yet well understood.

Factors in the environment that act through sensory and coding systems of the brain are more difficult to evaluate. A significant number of individuals who appear to have no evidence of preexisting brain pathology sometimes become psychotic in situations in which there is extreme sensory deprivation. At our present state of knowledge we cannot rule out undetected predisposing factors. In the first half of the 20th century the psychoses were often divided into organic and functional subtypes. It was assumed that the organic psychoses were caused by genetic factors or clear-cut brain pathology. The functional disorders were considered to be of unknown etiology, to be due to a pathology of brain function that was not associated with tissue damage, or to be primarily a response to pathological learning of coping skills or a response to certain types of stress. Because the term functional had all these connotations, the organic-functional distinction has become useless. Acute schizophrenia-like illness often appears in groups of individuals under extremely stressful conditions, such as military indoctrination training, religious seminary training, and wartime activities, but proving the etiological role of these experiences is very difficult. The chronic schizophrenia-like disorders are somewhat more common in the lower socioeconomic classes. This may suggest that the stress associated with poverty or lack of training in coping skills play a role, but it could just as well be that behavior among the lower classes is more likely to be identified as psychosis than it is among those with better social support systems. During the period from 1950 to 1975 it was common to talk of the role of family communication patterns in producing a predisposition to psychosis in children. Research has shown that the families of schizophrenics show patterns of within-family communication that differ from controls. In particular there is some evidence that schizophrenic families are less explicit or less open in communicating about intense emotions. It is difficult to come to a final conclusion at this time about the importance of *stress* and abnormal learning and psychological development as predisposing factors in specific types of psychosis. It is possible to conclude, however, that there is no single class of input factors playing a crucial etiological role for all episodes of behavior that can be called psychotic.

It may be more important to look at the systems of the brain that are involved in producing the normal behavior that becomes abnormal in psychosis. It is possible that a number of input factors of different categories all act in different ways on the same final common path somewhere in the complex systems of the brain that are the substrate for the higher cognitive functions.

Investigators have identified four ways of studying the brain mechanisms that are required for complex cognitive and emotional behavior and also for pathological behavior. In the first place, it is possible to infer plausible relevant brain mechanisms from the way in which input factors that are known to produce psychosis act on brain. For instance, because amphetamine and similar drugs can play a role in producing psychotic behavior, it is reasonable to explore the role of catecholamine transmitter systems in more detail. This has led to the so-called dopamine theory of schizophrenia pathogenesis. There is evidence supporting this theory that cannot be reviewed here, but our understanding of these systems is as yet very incomplete. Drug research may suggest specific biochemical, physiological, and anatomical investigations. Tumors and infections that are known to produce psychosis also can be shown through detailed neuropathological investigation to involve specific mechanisms of the nervous system. Unfortunately it is usually the case that nervous system tissue damage is so widespread that, although it helps identify regions of brain that are relevant, it does not pick out the precise system that is involved. By combining clues from pharmacology that give us some transmitter specificity and clues from neuropathology that identify key regions of brain that are involved, we may gain insight into the pathology of psychosis.

The second major approach is to begin with the categories of behavior that become abnormal in psychosis, then look for the brain mechanisms that are required for those normal functions, and finally to search for patterns of abnormal function in those systems that are associated with psychosis. Research on the brain mechanisms involved in attention, initiation of behavior, coordination of affective behavior with decision and optimization, concept formation, inferential reasoning, learning, memory, language, and many other functions may provide us with crucial clues.

A third approach to the explanation of psychosis is to follow up on the correlates of psychosis that we come upon accidentally. This correlational approach is very popular today, and it is probably due to the difficulty of following the first two approaches in any systematic fashion. Almost every possible way of measuring body processes and psychological function has been tried with psychotics, comparing the same individual in and out of the psychotic state, and comparing groups of psychotics with controls. The results of this research strategy have not been encouraging.

The last strategy used to develop an explanation of psychosis is to attempt a more systematic categorization of the heterogeneous class of psychotic episodes. We already know that it is possible to become psychotic under the influence of many different classes of determinants. If we subdivide the whole group into smaller subgroups, each of which is reasonably homogeneous, we may have a better chance to develop satisfactory explanations.

Further reading

American Psychiatric Association (1980): *Diagnostic and Statistical Manual of Mental Disorders*, 3rd ed.

Psychosurgery

Lyle W. Bivens

Psychosurgery is the surgical removal or destruction of brain tissue or the cutting of brain tissue to disconnect one part of the brain from another with the intent of altering behavior. Usually it is performed in the absence of direct evidence of existing structural disease or damage in the brain.

The surgical treatment of epilepsy, while in one sense a form of psychosurgery since behavioral symptoms are altered, should be excluded from this discussion when the disease can be clearly diagnosed and there is convincing evidence that epilepsy is caused by organic pathology in the brain. Of course any other neurosurgical treatment to repair or remove damaged brain tissue, or to remove tumors, is not psychosurgery.

In neuroanatomical terminology, the suffix *-tomy* refers to the destruction of brain tissue or the cutting of fibers in the brain. Lobotomy therefore is the destruction of tissue in the frontal lobes of the brain, or the cutting of fibers connecting the frontal lobes with the rest of the brain. Thalamotomy refers to the destruction of portions of the thalamus. The suffix *-ectomy* refers to the actual removal of tissue. Thus, a frontal lobectomy is the removal of the frontal lobes of the brain. Topectomy refers to selective removal of portions of the cortex, usually from the frontal region of the brain.

Ever since the first radical lobotomies were performed in which virtually all subcortical connections of the frontal lobe were severed, surgeons have attempted to refine the technique to limit the brain destruction in hope of reducing adverse side effects. Topectomy, which damaged less brain tissue than the early lobotomies, was tried. The development of undercutting techniques, whereby the subcortical connections of a selective part of the cortex are severed but the cortex itself is left intact, further refined the technique of topectomy.

The search for limitation of destruction of brain tissue (and presumably of adverse side effects) culminated in stereotaxic techniques, or the positioning of a small electrode in the brain. Geometric coordinates and x-ray inspection are used to place the electrode in a precise location, and then the tissue at the electrode tip is destroyed by passing a current through the electrode. With the development of stereotaxic techniques, structures deep within the brain become accessible for destruction, and amygdalotomy, thalamotomy, and hypothalamotomy began to replace lobotomy as psychosurgical procedures.

Psychosurgery was recommended for curing or ameliorating schizophrenia, depression, homosexuality, childhood behavior disorders, criminal behavior, and other psychiatric problems. For a period from about 1936 to 1955 lobotomy was enthusiastically embraced by a large number of medical and psychiatric practitioners in the United States and in England. The greatest proportion of psychosurgical treatments was upon schizophrenic patients, who were often difficult to manage and totally unable to function in society. Perhaps as many as 50,000 psychosurgical procedures were performed in the United States between 1936 and the mid-1950s.

The controversy over psychosurgery involves scientific, philosophical, political, and moral issues. In order to understand the nature and source of the psychosurgery controversy, it is necessary to make explicit some of the different viewpoints that are often unstated when the psychosurgery issue is discussed.

A fundamental concern about psychosurgery derives from different philosophical views of the relationship between mind (self) and brain. Much opposition to psychosurgery, and often the most vociferous opposition, is based on the conviction that any physical damage to the brain is tantamount to destruction of the "self." This viewpoint is most strongly illustrated by some of the rhetoric used by opponents of psychosurgery who equate it with "murder of the mind." Proponents of psychosurgery, while usually not articulating an alternative philosophy, do not equate the brain with the self and take a pragmatic approach to mental or behavioral disorders in which the primary criterion for selection of a treatment is whether it works.

A closely related issue is reflected in the different viewpoints about the causal factors in mental illness. Some psychosurgeons rationalize surgical treatment on the hypothesis that mental or behavioral disorders arise from biological dysfunction in the brain and that appropriate treatment must be based on manipulating or changing the biological substrate of behavior. Others, however, hold the view that disturbed behavior is a result of adverse environmental influences and that the solution to mental illness or behavioral disorders is to manipulate or change environmental variables. While both these views are extreme positions held only by a few, and are untenable in view of our current knowledge about the complex interrelations between environmental and biological causative factors, they illustrate another philosophical argument that, in frequently more subtle form than illustrated here, is one of the roots of the psychosurgery controversy.

Although virtually all psychosurgical procedures and technical innovations, including the first lobotomies, were suggested by experimental brain research with animals, the scientific rationale for any psychosurgical procedure is still quite tenuous. Generalizations from animal research have often been based on incomplete understanding of the complexity of behavior, logical deductions of dubious validity, and an uncritical acceptance of similarities of brain-behavior relationships in animals and humans. Although we know a great deal about how the brain influences a variety of specific and limited animal behaviors, our understanding of complex human emotional and cognitive behaviors is extremely limited. On the other hand, proponents of psychosurgery argue quite rightly that many medical therapies are based on a pragmatic criterion of effectiveness rather than an understanding of the physiological mechanisms underlying the disease or its treatment.

In contrast to most physical illnesses, many functional mental and behavioral disorders are poorly defined and difficult to diagnose, making the decision to treat with surgical means sometimes uncertain. Such problems also occur when judging the outcome of psychosurgical treatment, since the criteria for cure or ameliorization are not clear or universally agreed upon.

A key issue in the psychosurgery controversy is whether psychosurgery is an experimental procedure. Most psychosurgeons regard it as an accepted practice of proven efficacy, while critics see it as an experimental therapy based on alleged unpredictability of outcome, lack of evidence about efficacy, and lack of scientific rationale.

Alternative therapies to psychosurgery are another divisive issue. Although a great deal of research is being done on drug therapies and various forms of psychotherapy or behavior therapy, there are numerous instances in which none of these alternatives seems to offer any relief, and the patient is faced with a dehumanizing fate in an institution, often with pharmacologic restraints that equal or exceed any personality destruction that is claimed to be caused by psychosurgery. In these instances, psychosurgery might be seen as a reasonable last-resort therapy. On the other hand, there is no agreement or guidelines among practitioners about the duration, intensity, or degree to which other therapies should be tried before resorting to psychosurgery. Critics of psychosurgery claim, often correctly, that confinement in an institution is no guarantee of adequate attempts at therapeutic measures short of psychosurgery and that psychosurgery is frequently performed before other alternatives are sufficiently tried.

Closely related to the problem of psychiatric diagnosis is the extent to which mental or behavioral disorders are socially defined. This issue most often surfaces in the context of the psychosurgical treatment of aggressive or violent behavior in which critics of psychosurgery express the fear that it will be used for nefarious purposes as a means of controlling political or social dissidents. Stated in more general terms, critics charge that psychosurgery has been or can be used to change behavior for the convenience or comfort of persons other than the patient. Thus, there is claimed to be a bias toward the use of psychosurgery in blacks, women, and other minority or disadvantaged population groups.

Because of widespread public and congressional concern about psychosurgery in the mid-1970s, the U.S. Congress incorporated a mandate to study psychosurgery into the legislation establishing the National Commission for the Protection of Human Subjects of Biomedical and Behavioral Research. Research supported by the commission revealed that about 400 operations were being performed annually (for the period 1971–1973) by approximately 60 surgeons. The findings indicated that no significant psychological deficits are attributable to the psychosurgery in the patients evaluated and that psychosurgery was efficacious in more than half the cases studied. The data presented did not indicate that the procedure had been used for social control or that the procedure had been applied disproportionately to minority or disadvantaged populations. Specifically, it was reported from correspondence with the most active psychosurgeons in the United States that out of a combined total of 600 patients, 1 was black, 2 were Oriental Americans, and 6 were Hispanic Americans. Seven operations were reported to have been performed on children since 1970, and three prisoners underwent psychosurgery in Vacaville, California, in 1972. Most psychosurgery patients were middle class, had been referred to neurosurgeons by psychiatrists, and were about equally divided between male and female. More recent estimates of the frequency of psychosurgical procedures are that the annual rate is considerably lower now than in the period studied by the commission, due at least in part to the widespread publicity accorded the commission and its studies of the topic.

Probably the best-controlled study of psychosurgery is a continuation of a study begun by the national commission and currently supported by the National Institute of Mental Health. Suzanne Corkin and her colleagues at the Massachusetts Institute of Technology have been studying the long-term safety and efficacy of bilateral cingulotomy performed to treat severe pain or certain psychiatric disorders that have not responded to other treatments. A series of 179 patients were studied before and after surgery with a broad range of cognitive, personality, neurological, and psychiatric measures. Early results of this study have shown that this operation, which removes or destroys tissue by radiofrequency lesions in the cingulate gyrus (area 24 of Brodmann) and interrupts the cingulum bundle, has few if any undesirable side effects. The short-term results also indicate that patients treated for severe chronic pain were significantly improved on measures of psychiatric status and pain perception. Patients suffering from severe depression were also significantly improved, but obsessive-compulsive or schizophrenic patients did not improve on a six-month postoperative measure of global mental health.

Further reading

Corkin S, Twitchell TE, Sullivan EV (1979): Safety and efficacy of cingulotomy for pain and psychiatric disorder. In: *Modern Concepts of Psychosurgery*, Hitchcock et al, eds. Amsterdam: Elsevier

National Commission for the Protection of Human Subjects of Biomedical and Behavioral Research (1979): *Report and Recommendations: Psychosurgery*. Department of Health and Human Services, Pub No (OS) 77–002. Washington DC: US Government Printing Office

Valenstein ES (1973): *Brain Control: A Critical Examination of Brain Stimulation and Psychosurgery*. New York: Wiley

Valenstein ES, ed. (1980): *The Psychosurgery Debate: Scientific, Legal, and Ethical Perspectives*. San Francisco: WF Freeman

Schizophrenia

Erik Strömgren

When in 1908 Eugen Bleuler introduced the term schizophrenia, he did not intend to create a new concept. He just wanted to suggest a new term for the concept of dementia praecox, which had been developed by Emil Kraepelin during the preceding 15 years and had received widespread acceptance. Kraepelin's demonstration that dementia praecox and manic-depressive illness constituted two groups of disorders that were basically different from all other psychoses and from each other helped to clarify a clinical field that until then had been more or less chaotic. As late as the middle of the 19th century many psychiatrists believed that there was only one mental disorder, and that the different clinical forms merely represented different stages of the same disease. Gradually, however, it became clear that those psychoses that could be demonstrated to be caused by a physical morbid process had characteristic psychiatric syndromes the presence of which would always indicate that the disorder was somatically determined: in acute stages, disorders of consciousness and delirium; in chronic stages, dementia, deterioration of memory and other intellectual functions. In the majority of psychoses, however, no such "organic" symptoms were present, and there were no signs of physical effects on the brain. These psychoses were called, in distinction to the "organic" group, "functional" or "endogenous." Kraepelin's longitudinal studies on functional psychoses showed a tight correlation between symptomatology and course: when the symptoms were mainly affective, in the form of mania or depression, the prognosis was in principle good. On the other hand, if the symptomatology was dominated by delusions, hallucinations, withdrawal, emotional blunting, odd and incomprehensible behavior and speech, the prognosis in most cases was bad. These observations led to the distinction between manic-depressive disorder and dementia praecox. In the latter disorder the symptomatology varied from case to case, and in the same patient over time. Kraepelin distinguished four clinical types: hebephrenia, characterized mainly by withdrawal and emotional blunting; paranoid form, dominated by delusions; catatonia, characterized by disturbances in attitudes and movements; and dementia simplex, characterized by arrest of most mental functions. The distinction between hebephrenia and dementia simplex was, however, not very clear.

In 1911 Eugen Bleuler published his classic monograph on dementia praecox or the group of schizophrenias. The reason why Bleuler found the term dementia praecox inadequate was that these patients were not "demented"; their intellectual faculties seemed unchanged; further, the disease did not always start "praecociter," i.e., in youth. The term "schizophrenia" (splitting of mind), on the other hand, pointed to fundamental features of the symptomatology. The title of the book stressed that Bleuler did not regard schizophrenia as a nosological entity; several pathogenic agents might lead to the same clinical picture. Although Bleuler mentioned that his contribution was founded on the ideas of Kraepelin, he added many important observations and ideas to the understanding of schizophrenics. First of all, he distinguished between "basic" symptoms and "accessory" symptoms, respectively (the terms "primary" and "secondary" symptoms have also been used). Bleuler, like Kraepelin, regarded the etiology of schizophrenia as organic, the basic symptoms being caused directly by the organic brain process, namely, disturbances of the association processes and emotional reactions and "autistic" disturbances of contact with other human beings. All other symptoms—delusions, hallucinations, catatonic symptoms, etc.—he regarded as psychologically understandable reactions to the presence of the basic symptoms. The distinction between basic and accessory symptoms was strengthened by the observation that the basic symptoms never disappeared, but accessory symptoms were often subject to change and might disappear intermittently or permanently.

Although Bleuler's delimitation of schizophrenia was very similar to Kraepelin's delimitation of dementia praecox, he did add descriptions of some mild cases of schizophrenia that Kraepelin would probably not have included in his concept. Both investigators stressed the bad prognosis of the disorder, although in a small percentage recovery seemed to occur.

Since the days of Kraepelin and Bleuler the question of the delimitation of schizophrenia has been the subject of endless discussions, some psychiatrists advocating a very narrow delimitation, others being inclined to include the great majority of nonorganic psychoses within the concept. A number of symptoms are, however, by all psychiatrists regarded as highly indicative of a schizophrenic process: delusions, mainly of persecution, hallucinations, mainly acoustic, disorders of thought, unusual and incomprehensible associations, inadequacy of emotional reactions, odd and inadequate behavior, emotional and behavioral withdrawal, loss of sense of reality. Most of these symptoms may occur also in other disorders and can therefore not be regarded as pathognomonic. Certain combinations of such symptoms are, however, seriously indicative of schizophrenia. Much attention has been paid to some of the symptomatological criteria described by Kurt Schneider as essential for the diagnosis of schizophrenia. Among symptoms that appear in schizophrenia he suggests that some are of the "first rank." These symptoms have empirically turned out to be of special importance when psychiatrists make a diagnosis of schizophrenia. They are the following: patient hearing his own thoughts spoken aloud; hearing voices talking to each other about the patient; hearing voices which comment on the patient's behavior; feelings of patient's body being influenced by outside forces; feelings of thoughts being removed or foreign thoughts being inserted into the mind; feeling that thoughts are broadcast to other people; feeling that emotions, drives, and intentions are dictated by external forces. Schneider concludes: "When it is stated that a patient has such experiences, and there is no evidence of a physical disorder, we in all modesty are talking about schizophrenia."

What is essential is, in any case, that these symptoms are present in spite of clear consciousness. If the "schizophrenic" symptoms occur in patients with delirium or other disturbances of consciousness, they are in no way indicative of schizophrenia.

Etiology

We have no reason to believe that schizophrenia is a nosological entity, and there is no basis for singling out special subgroups of schizophrenia as belonging to specific etiological entities. Most explorations into etiology have therefore been based on heterogeneous patient groups. This explains the diversity of results.

In the beginning of the 20th century it was generally accepted that the etiology of schizophrenia was mainly hereditary. The first systematic studies of twins supported this viewpoint, concordance among monozygotic twins being about 75%. The results of a number of later studies, however, indicate that the concordance is about 40%, regardless of whether the twins have been living together or have been separated at early age. Although heredity thus plays an important role in the origin of schizophrenia (which has also been demonstrated by family and adoption studies), it is clear that the greater part of the etiology must be exogenous. Whether these exogenous factors are mainly physical or psychosocial has yet to be determined. Some physical agents can give rise to psychoses that are clinically indistinguishable from schizophrenia. A number of drugs (amphetamine, other stimulants, marijuana, etc.) can give rise to such schizophrenia-like psychoses, which in most cases disappear when the drug is no longer available. A number of brain disorders (Huntington's chorea and encephalitis) may have a similar action that is of special interest since the anatomical site of the underlying processes is well known. Speculations concerning the possible site of a supposed schizophrenic process are therefore not unrealistic. Although a number of organic etiologies to schizophrenia-like psychoses are thus well known, it should be remembered that only in a small percentage of all schizophrenias can such physical factors be ascertained. Anatomical, biochemical, and pharmacologic studies of the brains of schizophrenics have given controversial results. Some investigators have claimed to have disclosed biochemical abnormalities in the brains of schizophrenics; other researchers have been unable to reproduce the results. It is obvious that it is difficult to carry out such studies and to collect data, mainly because untreated schizophrenics are very rare. Changes in the brains of schizophrenics who have received different treatments known to cause changes in the biochemistry of the brain give inconclusive results. A further problem is the possibility that a schizophrenic process might be intermittent; the biochemical findings would then be different in active and inactive phases of the process. At present no biochemical process has been demonstrated in the brains of schizophrenics that could distinguish them from normal persons or from other psychotics.

Anatomical findings are no more conclusive than the biochemical findings. Claims have repeatedly been made that aerograms and computerized tomography (CT) scanning have disclosed areas of atrophy in the brains of schizophrenics. The importance of such findings is, however, dubious since it has become clear that reductions in brain mass can be reversed and that they may be caused by drug treatment. Further research that takes into account the different sources of error is urgently needed.

When the psychopharmacological era started, and spectacular therapeutic results were obtained, it was natural to assume that the biochemical changes occurring during such treatment (with regard to biogenic amines, transmitters, etc.) could lead to the detection of biochemical processes that might be of etiological importance for schizophrenia. Little insight has been gained in this way, however, one reason being that drugs that can help schizophrenics have therapeutic impact on other disorders, so the biochemical findings made cannot possibly be specific for schizophrenia. The situation is much more complicated than in manic-depressive disorder, where at least one drug (lithium) has a specific therapeutic relationship to the disorder, and where studies on the effect of that drug are therefore much more promising.

Although no physical factors have been demonstrated that are essential to the etiology of schizophrenia, it would be naive to assume that such factors do not exist. It is, on the other hand, natural that the inconclusive results of the search for physical etiological factors could stimulate the interest in the search for psychosocial causes. Here again the results are, unfortunately, meager. It has not been possible to demonstrate factors that are of convincing importance for the onset of schizophrenia. In most cases it seems impossible to understand why the schizophrenia started at all, and why it started at a particular moment. The premorbid situations of schizophrenics are so different that no generalizations are possible. It is amazing that schizophrenia seems to have practically the same incidence in all populations studied. Such studies have been performed in populations with different cultures, political structures, and socioeconomic levels. Differences in prevalence of schizophrenia among different parts of the populations seem primarily to have been caused by a drift of the schizophrenics to special sections of the populations. The socioeconomic level of parents of schizophrenics is average.

The ubiquity of schizophrenia in all human populations, regardless of the immense diversity of environmental conditions, would seem to indicate that noxious factors relevant to the development of schizophrenia must be active in situations common to all human beings. Focus has been on the emotional relationship between children and parents (or other emotional key persons). In many cases of schizophrenia, infancy and early childhood relationships have been shown to be of a most noxious nature. In the majority of cases, however, such situations have not been ascertained. Manfred Bleuler, who has performed the most intensive personal studies of schizophrenics, found that many of his patients had been subject to severe stress in childhood; the nature of the stress was, however, very different, and generalizations with regard to pointing out certain conflicts as essential could not be made.

The basis for formulating a theory concerning the etiology of schizophrenia is very weak. However, based on what is known with reasonable certainty today, the following summary statement may be permissible:

A genetically determined vulnerability seems to play a role in the origin of schizophrenia. This genetic basis differs from case to case, certainly quantitatively and possibly qualitatively. Early acquisition of minimal brain damage may enhance the vulnerability. During early childhood these vulnerable individuals may, in reaction to frustrations in the emotional relationship to their peers, develop an ambivalent fear of establishing emotional contact, which enhances their vulnerability to frustrations and their tendency to react with withdrawal from emotional attachment, and from reality as well, thus developing autism that seems to be the central, maybe pathognomonic feature of schizophrenia. From case to case there are probably great differences with regard to the relative impact of genetic, somatic, and psychological factors.

Therapy

The possibilities for therapy of schizophrenia have improved radically during recent decades. Before the potent physical therapies were introduced, it was known that the course of schizophrenia in most cases depended on environmental factors such as mental stimulation and occupational therapy. Such factors are essential even in the era of somatic therapies. The trend is to begin treatment with psychotropic drugs and supportive psychotherapy, usually starting during short hospitalization and continuing within the framework of a community mental health setup. In most cases improvements can be observed, including social and familial adaptation. In very few cases, however, do schizophrenics become symptom-free; at least some degree of autism is left.

Relationship to other groups of disorders

Although there is a core group that would be diagnosed as schizophrenics by the majority of psychiatrists, there are other more peripheral groups included in schizophrenia by some psychiatrists and regarded as separate groups by others. This is especially true for patients who are predominantly paranoid and show little or no other schizophrenic symptomatology. Paranoid psychoses arising in elderly persons and associated with hallucinations, but with well-preserved personality and little autism, are sometimes singled out as "paraphrenia." Some "borderline states" of pseudoneurotic or psychopathic schizophrenia have been shown to have close genetic relationships to schizophrenia. Schizophrenia-like psychoses in epileptics have, on the other hand, been demonstrated not to have any genetic relationship to schizophrenia. Schizophrenia-like psychoses arising in connection with severe, acute mental stress are likewise without genetic relationships to schizophrenia. Many schools of psychiatry label these as "reactive" or "psychogenic" psychoses.

International nomenclatures and classifications

The International Classification of Diseases of the World Health Organization (ICD, 9th edition) introduces section 295 "schizophrenic psychoses," with a detailed description of the type of psychoses included under this heading, a description that can no doubt be regarded as applicable to the great majority of schizophrenic cases. Section 295 comprises a number of subgroups that should naturally be included under schizophrenic psychoses: simple type, hebephrenic type, catatonic type, paranoid type, latent schizophrenia, and residual schizophrenia. There are two more controversial subgroups. In "acute schizophrenic episode" there is a dreamlike state with slight clouding of consciousness. Such cases would not be included in schizophrenia by most psychiatrists, clear consciousness being regarded as one of the features essential for the diagnosis of schizophrenia. In "schizoaffective type," there is a mixture of schizophrenia-like features, and many psychiatrists regard these cases either as a separate group or as a subgroup of affective psychoses.

The *Diagnostic and Statistical Manual of Mental Disorders* (DSM III), which appeared in 1980, introduced a radical reclassification of schizophrenia, distinguishing between "schizophrenic disorders," "paranoid disorders," and "psychotic disorders not elsewhere classified," the latter group comprising the subgroups "schizophreniform disorder," "brief reactive psychosis," and "atypical psychosis." Schizophreniform disorders are said to have symptomatology identical with that of schizophrenia, with the qualification that the duration must be less than 6 months but more than 2 weeks. If the disorder lasts for less than 2 weeks, it is called a "brief reactive psychosis"; in addition, the psychosis must follow a psychosocial stressor; if such stressor is not present, the case is called "atypical psychosis." If a schizophrenia-like psychosis arises after middle age, it is also labeled atypical psychosis. It is obvious that there is a radical and disturbing discrepancy between DSM III and ICD 9 in these respects. DSM III contains a further innovation, the schizotypal personality disorder, which is not regarded as a form of schizophrenia; many psychiatrists feel that these cases should be classified as schizophrenia, pseudoneurotic or pseudopsychopathic. Such cases seem to have a genetic relationship to schizophrenic psychoses and may be included in the so-called "schizophrenic spectrum."

Diagnostic criteria

Since the delimitation of the concept of schizophrenia has varied between different psychiatric schools, statistics on the incidence of schizophrenia have been difficult to compare. The differences have also hampered research and comparisons of therapeutic results. It is therefore natural that there has been interest in establishing recognized diagnostic criteria that could secure that groups of schizophrenics that are compared really are comparable. In recent years a number of systems of "research diagnostic criteria" (RDC) have appeared. It is obvious that groups delimited by means of the same RDC are more comparable than groups described by means of subjective clinical criteria that have never been standardized. On the other hand, there is no basis for the contention that by means of such RDCs, which contain relatively few items, any homogeneous groups or nosological entities can be delimited. Diagnostic criteria are adequate for differentiation of two different groups; they are less useful in delimiting one homogeneous group. For this purpose much more differentiated and validated interview schedules are necessary, for instance in the form of the *Present State Examination* (PSE) devised by J.K. Wing. The *PSE* has been widely accepted as a useful tool, not least within the cross-national investigations organized by the World Health Organization, for instance, in the *International Pilot Study of Schizophrenia* (IPSS).

Further reading

Bleuler E (1911): Dementia praecox oder Gruppe der Schizophrenien. In: *Handbuch der Psychiatrie. Spezieller Teil, 4. Abteilung, 1. Hälfte*, Aschaffenburg G, ed. Leipzig and Wien: Franz Deuticke. English ed. (1950): *Dementia Praecox or the Group of Schizophrenias*. New York: International Universities Press

Bleuler M (1972): *Die schizophrenen Geistesstörungen*. Stuttgart: Georg Thieme Verlag. English ed. (1978): *The Schizophrenic Disorders*. New Haven and London: Yale University Press

Neale JM, Oltmanns TF (1980): *Schizophrenia*. New York, Chichester, Brisbane and Toronto: John Wiley & Sons

Wing JK, Wing L, eds. (1982): *Handbook of Psychiatry. Vol 3: Psychoses of Uncertain Aetiology*. Cambridge, London, New York, New Rochelle, Melbourne and Sydney: Cambridge University Press

Sleep Disorders

Thomas Roth, Timothy Roehrs, and Frank Zorick

Sleep disorders medicine has become an important clinical and research discipline for two reasons. The first is the documentation of sleep-wake complaints. A recent national survey found that one-third of the American population reported some degree of insomnia, with 17% of the population considering their insomnia serious. In addition, up to 6% of the population complain of excessive sleepiness during the day. The second and more immediate impetus to sleep disorders medicine was the discovery of sleep-specific pathologies. That is, individuals who show normal physiological functioning during the neural state of wake can show significant pathology when in one of the two neural states of sleep (REM, or rapid eye movement, and NREM, or non-REM). In addition, in perfectly healthy individuals, control of basic physiological functions such as respiratory drive, cardiac rhythm, and thermoregulation have different laws during wake and sleep. In fact, control of these functions varies within the two neural states that constitute sleep.

Historically, sleep-wake complaints in the population were dealt with at a superficial level. Insomnia typically was viewed as a single entity primarily caused by psychological-psychiatric conditions. On the other hand, daytime sleepiness was synonymous with narcolepsy. That is, patients who complained of excessive daytime sleepiness regardless of their signs and symptoms were invariably diagnosed as narcoleptic. In fact, it is interesting to note that since patients with daytime somnolence often exhibit different signs and symptoms, different types of narcolepsy were postulated (e.g., REM narcolepsy and NREM narcolepsy). With increased research interest in sleep disorders and the development of standardized methods for the all-night evaluation of physiological functioning of these patients, it soon became evident that a host of different abnormal physiological events (e.g., periodic leg movements, apneas, alpha intrusions) occur in these patients' sleep, and these events can lead to sleep disturbance and, hence, sleep-wake complaints.

The diagnostic classification system of sleep-wake disorders developed in 1978 advanced our understanding of sleep disorders. The intent of this system was to provide clinicians with a listing of the signs and symptoms of the various sleep disorders, as well as their differential diagnoses. The systematic use by sleep clinicians of this diagnostic system over the past six years has led to three major conceptualizations about the nature of sleep disorders.

First, it emphasized that insomnia and daytime sleepiness are symptoms, not specific diseases entities. As with any other symptom in medicine, the severity and duration of the symptom is important in assessing its significance, diagnosis, and treatment. In assessing severity, it must be recognized that although a patient may complain severely, the complaints may be associated with no objective evidence of any sleep disturbance. It is essential that patients who complain of insomnia show signs of a sleep disturbance and patients complaining of daytime somnolence show evidence of increased daytime sleep tendency before treatment is instituted.

The development of all-night sleep recording (clinical polysomnography) and repeated test of daytime sleep tendency (multiple sleep latency test) enables clinicians and researchers to validate and measure the presence and severity of sleep complaints. In the absence of positive findings on these tests, it is inappropriate to treat a patient for a sleep disorder. This is not to say that patients with sleep complaints but no objective evidence of a sleep disorder should be considered as malingerers or simply dismissed. Rather these patients' symptoms, and hence their treatments, are not related to abnormalities of sleep-wake mechanisms.

The duration of the symptom is equally important. Generally in medicine, symptomatic treatment is reserved for transitory problems. A recent National Consensus Conference on Drugs and Insomnia concluded that the same is true for the symptom of insomnia. While hypnotics are consistently effective and, thus, indicated in transient and short-term insomnia, they are not recommended as the standard primary treatment in patients with chronic insomnia. In evaluating daytime sleepiness, duration of the problem is similarly important. When extreme daytime sleepiness is present for short periods of time, acute sleep deprivation is typically the cause. However, when the sleepiness is chronic, sleep loss per se accounts for only a small proportion of patients who suffer from daytime sleepiness.

The second major concept to be derived from the diagnostic classification system is that sleep disturbances vary in type and etiology. Some patients exhibit sleep disturbances characterized by prolonged periods of wakefulness. This type of disturbance is most pronounced in patients with insomnia associated with depression or restless legs. These two categories of patients show a most pronounced sleep disturbance in terms of increased wakefulness during the night. Another type of sleep disturbance is characterized by multiple brief arousals during the night. This disturbance typically is seen in patients with nocturnal myoclonus or apneas during sleep. A third type of sleep disturbance is characterized not by a change in the continuity of sleep, but by the superimposition of wake electroencephalographic activity (EEG) on the sleep EEG. This finding, alpha-delta sleep, is seen in patients with fibromyositis syndrome or some depressions. It is interesting that such patients do not have typical insomnia complaints (i.e., difficulty falling asleep or frequent nocturnal awakenings), but rather they complain about the quality of their sleep. They report no sleep, light sleep, or unrefreshing sleep. Finally, some patients show a sleep disturbance characterized by normal sleep, which because it is out of phase with a person's bedtimes, results in a prolonged period of wakefulness following or preceding the normal sleep. These are circadian rhythm disorders where there is a mismatch between the person's sleep-wake rhythm and his bedtimes. Acutely, this can be seen in people experiencing jet lag or shift work, and chronically in people with phase delay or phase advance syndromes.

These disturbances of sleep can be caused by primary sleep pathologies, by waking medical or psychiatric disorders, by external agents such as drugs, or by environmental factors. All waking medical diseases continue to affect the patient when he falls asleep and, thus, have the potential to cause sleep disturbances. For example, any disorder associated with pain or discomfort (e.g., reflux) during wakefulness has the potential to cause repeated arousals and awakenings during sleep. In addition, since the control of physiological functions differs in sleep and wakefulness, some disorders become exacerbated as the patient goes from waking to sleeping. Many patients with waking lung disease experience exacerbation of their condition while asleep. In some this is more pronounced in REM sleep than in NREM sleep owing to the differential control of respiration in these two sleep states. Waking psychiatric diseases, as noted earlier, are often associated with disturbed sleep. One of the hallmark symptoms of endogenous depression is disturbed nocturnal sleep, especially early morning awakenings. Drugs can also lead to sleep disturbances. It is obvious to most individuals that stimulants such as amphetamines, methylphenidate, and caffeine can cause disturbance of sleep. However, other drugs such as hormones (e.g., cortisone, progesterone), respiratory stimulants (e.g., theophylline), and autonomic agents cause sleep disturbance as well. Finally, drugs used to promote sleep are typically general central nervous system (CNS) depressants. Yet, it is critical to recognize that these drugs have the potential to disturb sleep. Alcohol is the best example of this. Despite the fact that alcohol hastens sleep onset, it has been shown, in both human and animal studies, to lead to severe sleep disturbance. In fact, a review of the sleep-pharmacology literature indicates that virtually all classes of drugs that promote sleep can in certain situations lead to sleep disturbances. Finally, sleep disturbances can be found in drug-free individuals who are perfectly healthy when awake, but suffer from primary sleep disorders. Apneas during sleep and nocturnal myoclonus are the clearest examples of primary sleep disorders. These physiological abnormalities may be totally occult while the person is awake (except for a sleep-wake symptom) and can only be detected by evaluating physiological functioning during sleep.

The third major concept to be derived from the diagnostic classification system and its use deals with the relation of sleep-wake complaints to sleep disturbance. The diagnostic classification system is a symptom-oriented system. Its two major divisions list the disorders associated with the symptoms of insomnia and the disorders that lead to symptoms of daytime sleepiness. With the exception of narcolepsy, every major disorder appears on both listings. What this means is that there are multiple etiologies that can cause sleep disturbances, and each of these can in turn lead to a complaint of insomnia or a complaint of daytime sleepiness. There are two possible explanations for this. First, all patients have comparable sleep disturbances, but some patients focus their attention on their sleep, and hence complain of insomnia, while other patients focus their attention on their waking function, and hence complain of daytime sleepiness. An alternative explanation is that the subjective experience of impaired sleep and the subjective experience of impaired alertness are two points on a continuum of sleep fragmentation. That is, infrequent arousals during the night lead to long periods of wakefulness and subjective complaints of insomnia. Frequent arousals lead to an accumulation of a sleep debt, which in turn leads to brief arousals and a complaint of daytime somnolence. This explanation seems more plausible as studies in several diagnostic groups including nocturnal myoclonus and apnea show that patients complaining of daytime sleepiness show more frequent and briefer arousals than patients complaining of insomnia. Regardless of the explanation, the point is that a single sleep disorder may result in a variety of sleep-wake complaints or even no complaints. Thus, unless one wants to treat patients purely on a symptomatic level (hypnotics for insomnia and stimulants for daytime sleepiness), the focus should not be solely on the symptom. More appropriately, the specific sleep disturbance must be recognized, understood, and treated.

This brief review of the nature of sleep disorders emphasizes the increased understanding of sleep disorders at a clinical level. However, the understanding of the pathophysiology of these disorders has not kept pace with the ability to identify them clinically. A survey of current treatments for the various sleep disorders makes it clear that more research into the pathophysiology underlying these disorders is needed if the pathophysiology is to be reversed. All primary sleep disorders are, by definition, central nervous disorders. That is, there is a certain level of functioning in wake, and with a change in neural state (sleep) there is a change in physiological functioning. For example, the vast majority of patients with sleep apnea syndrome show normal respiratory function while awake, but frequent airway obstructions while asleep. In fact, during REM sleep, the hemodynamic consequences of these obstructions are more pronounced than during NREM sleep. Of the currently available treatments for apnea, tracheostomy provides an artificial airway below the site of obstruction, continuous positive airway pressure attempts to reverse the negative pressure caused by inspiratory effort, and other surgical treatments (uvulopalatopharyngoplasty and mandibular advancement) attempt to increase airway size. None of these are intended to reverse the CNS defect in apnea syndrome. Similarly in nocturnal myoclonus, no medications have been shown to inhibit these muscle bursts, and all currently effective treatments blunt nonspecifically the CNS arousal associated with these bursts. Narcolepsy, which has been recognized as a CNS pathology for over one hundred years, can still only be managed symptomatically by the use of stimulants. In summary, sleep disorders medicine has made major strides in terms of the identification and clinical management of these disorders. The future goal for this field lies in the identification and reversal of the underlying pathophysiology of these disorders. The realization of this goal is dependent on future research on the neurophysiology and neuropathology of sleep.

Further reading

Association of Sleep Disorders Centers (1979): *Diagnostic Classification of Sleep and Arousal Disorders,* prepared by the Sleep Disorders Classification Committee, Roffwarg HP, chairman. *Sleep* 2:1–37

Orem J, Barnes C (1980): *Physiology in Sleep.* New York: Academic Press

Guilleminault C (1982): *Sleeping and Waking Disorders: Indications and Techniques.* Menlo Park, Calif: Addison-Wesley

Hauri P (1982): *The Sleep Disorders. Current Concepts.* Kalamazoo, Mich: Upjohn Company

Stress, Neurochemistry of

Adrian J. Dunn

Most neurochemical parameters are altered during or following some form of experimental stress. But, because all stressful treatments (stressors) are complex, involving a variety of stimuli, it is difficult to be certain that the responses are associated unequivocally with stress per se. While the concept of stress is generally understood, there is no generally accepted definition. Stress is subjective; what is stressful for one person may not be so for another, and even what is stressful for one person on one occasion may not be so on another. Thus, most definitions of stress cite the person's response, the most widely accepted being the activation of the pituitary-adrenal axis. This definition derives from the work of Selye, who discovered the adrenocortical activation in stress and promulgated the theory that stress is a nonspecific (adrenocortical) response. Surprisingly, his theory and the definition that arises from it ignore the stress-related activation of the adrenal medulla and the sympathetic nervous system, established by the pioneering work of Cannon. Most modern researchers accept the ubiquity of the adrenocortical response in stress, but would not agree that the stress response is physiologically nonspecific (for example, the physiological response to cold exposure is not the same as that to heat).

Activation of the adrenal medulla results in secretion of epinephrine and, in most species, norepinephrine (NE) into the general circulation. The sympathetic nervous system contributes NE both locally and into the general circulation. Activation of the hypothalamic-pituitary-adrenal (HPA) axis results in the secretion of corticotropin-releasing factor (CRF) from the hypothalamus, adrenocorticotropin (ACTH) from the pituitary, and glucocorticosteroids (cortisol and/or corticosterone, depending on species) from the adrenal cortex (see Fig. 1). β-Lipotropin β-LPH) and small amounts of β-endorphin are also released concomitantly from the anterior pituitary. Physiologically, the common responses to all stressors appear to be the activation of the sympathetic nervous system (considered to include the adrenal medulla as argued by Cannon), and the activation of the HPA axis. This results in elevation of the plasma, epinephrine, and NE, as well as ACTH and glucocorticoids. Stressors also activate cerebral noradrenergic neurons, which can perhaps be regarded as the counterpart in the central nervous system (CNS) of the peripheral sympathetic nervous system.

Activation of these two systems (the entire sympathetic and the HPA) comprises the core of the stress response; however, the systems are not independent (Fig. 2).

There is extensive evidence that catecholamines are involved in the regulation of HPA activity; noradrenergic neurons are thought to inhibit CRF release via a hypothalamic α-receptor, while in rodent pituitary a β-receptor activates the release of ACTH. CRF and ACTH affect CNS catecholaminergic systems (generally an activation). Also, glucocorticoids exert considerable control over catecholaminergic systems in both the CNS and the periphery. Most notably, corticosterone increases the activity of adrenal medullary phenylethanolamine-*N*-methyltransferase (PNMT), the enzyme that converts NE to epinephrine, hence increasing the ratio of epinephrine to NE secretion from that gland. But glucocorticoids also play a permissive role in the adaptation of adrenergic receptors, both in the brain and in the periphery. Other neurochemical responses to stress may be secondary to any of these.

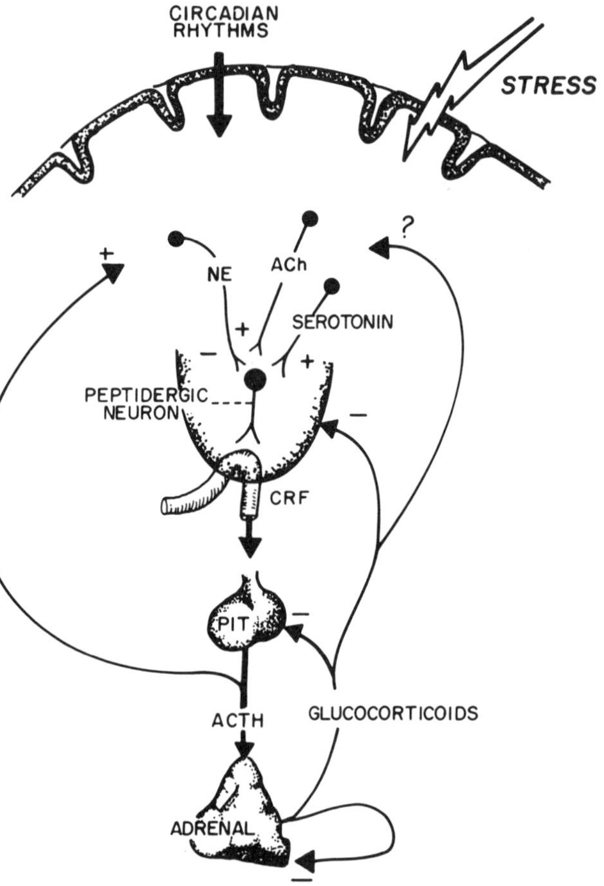

Figure 1. The hypothalamic-pituitary-adrenocortical (HPA) system. Noradrenergic (NE), cholinergic (ACh), and serotonergic inputs to the hypothalamus alter the release of corticotropin-releasing factor (CRF), which stimulates release of ACTH from the anterior pituitary (PIT), which in turn stimulates release of glucocorticoids from the adrenal cortex. Glucocorticoid feedback occurs at all levels, whereas ACTH feedback may occur on the brain. From Dunn and Kramarcy (1984).

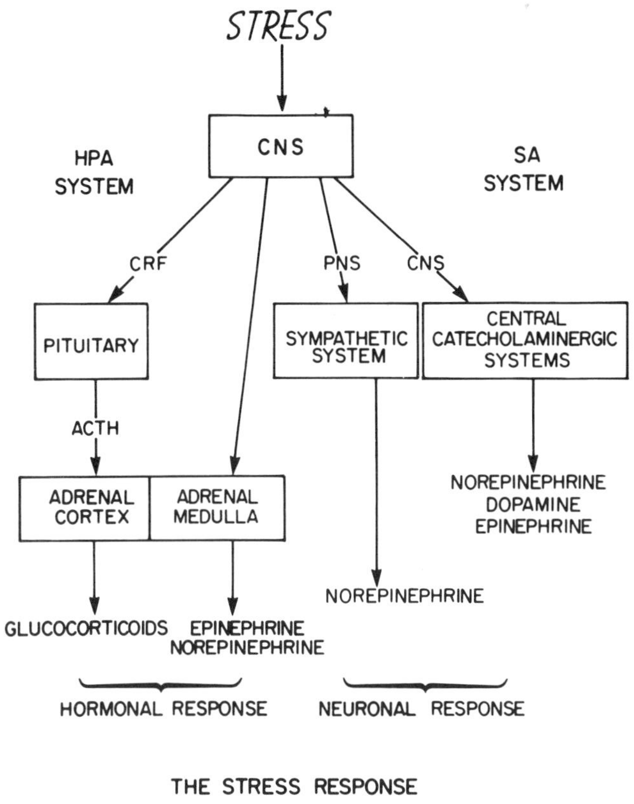

Figure 2. A schematic indicating the overall arrangement of the hypothalamic-pituitary-adrenocortical (HPA) and sympathoadrenomedullary (SA) systems. The central noradrenergic system appears to be activated in parallel with the peripheral sympathetic nervous system and can be considered the counterpart in the central nervous system of that system. The other cerebral catecholaminergic systems (i.e., dopamine and epinephrine) may also be activated in stress. The diagram illustrates the parallel activation of both the HPA and SA systems and their interaction in the adrenal gland. From Dunn and Kramarcy (1984).

The catecholamine responses

Numerous studies have shown that stressful treatments can decrease cerebral NE content. This effect presumably reflects increased release of NE coupled with the inability of synthetic mechanisms to keep pace, because it occurs only after more intense or prolonged stress. Electrophysiological recording and studies of biochemical turnover provide more sensitive measures of noradrenergic activity and can detect activation of cerebral noradrenergic neurons following briefer and milder stressful treatments. Changes of NE turnover can be estimated by following its disappearance after blockade of synthesis with drugs such as α-methyl-*p*-tyrosine. But, because such drugs themselves activate the HPA axis, measurement of the production of catabolites, such as 3-methoxy,4-hydroxyphenylethyleneglycol (MHPG) or its sulfate derivative is preferred. This activation of cerebral NE neurons seems to be general in stress, and occurs under the same conditions as activation of the sympathetic nervous system. The noradrenergic response seems to be regionally nonspecific, in keeping with the anatomical and biochemical data indicating that NE cell bodies in the locus coeruleus send collaterals to widespread areas of the brain.

Extensive data now indicate that certain of the cerebral dopamine (DA) systems are also activated during stress. This applies particularly to the mesocortical DA system which projects to prefrontal and cingulate cortices, but some data implicate other regions, e.g., the nucleus accumbens, and the response may be more general. This DA response is readily detected in frontal cortex by assay of turnover, e.g., the production of the catabolite, 3,4-dihydroxyphenylacetic acid (DOPAC). It occurs following a variety of stressors, including psychological (conditioned) stress. Other data suggest that cerebral epinephrine-containing neurons are also activated during stress, although the generality of this response remains to be determined. Thus it may be that neurons containing all three catecholamines in the brain contribute to the stress response.

The data for an involvement of cerebral serotonin (5-HT) in stress are less clear. While numerous articles indicate alterations of 5-HT content or metabolism in stress (especially increases in the production of serotonin's major catabolite, 5-hydroxyindolacetic acid, 5-HIAA), in most cases very severe treatments have been used. Some evidence is consistent with the possibility that the changes in cerebral 5-HT are secondary to increases of circulating corticosterone.

A few data exist to suggest effects of stress on other neurotransmitters. These especially include acetylcholine and gamma-aminobutyric acid (GABA), but changes in the metabolism of these seem to be less sensitive measures of stress. The changes in GABA are particularly interesting because of the intimate relationship between this neurotransmitter and the benzodiazepines (the so-called minor tranquilizers), which find extensive clinical use for their antianxiety (i.e., anxiolytic) effects.

Neurochemical effects secondary to the activations of catecholaminergic systems include an activation of adenylate cyclase and consequent increases in cyclic AMP. Chronic stress increases the activity of the biosynthetic enzymes for catecholamines, especially in the sympathetic nervous system, but also in the brain. These occur largely as a result of increased neuronal activity, but glucocorticoids appear to be important regulators and, in some cases, mediators. There also appear to be a desensitization of adrenergic receptors, indicated both by a decreased number of binding sites for β-adrenergic ligands and a decreased responsivity of adenylate cyclase to NE. These changes resemble those that occur following chronic treatment with agonist drugs, in particular the tricyclic antidepressants.

Activation of the HPA axis

Activation of the HPA axis results in the secretion of CRF, ACTH, and glucocorticoids into the circulation. Thus, secondary effects may occur from any of these hormones, or even from β-endorphin, which may be released concomitantly with ACTH from the pituitary during stress. Some of these effects may reflect feedback mechanisms involved in the regulation of the HPA axis, but others are clearly independent. Glucocorticoid administration to rats early in life can cause long-lasting impairments of brain weight, cell proliferation, dendritic branching, and myelination. There is good evidence that glucocorticoids may play a role in the normal regulation of myelination. Thus glucocorticoids may mediate the adverse effects of stress in immature or even fetal animals. Glucocorticoids have also been implicated in age-related changes in the brain. Clinically, glucocorticoids are used extensively to reduce vasogenic brain edema, reflecting a role in ion and water transport. Well-documented effects of glucocorticoids on the brain include induction of glycerol phosphate dehydrogenase, an enzyme intimately involved in lipid metabolism, and alterations in the metabolism of several neurotransmitters, especially the catecholamines and 5-HT, but also acetylcholine and GABA.

Effects secondary to ACTH release may include a small general increase of protein and perhaps RNA synthesis, increases in cyclic AMP and specific phosphoproteins, increases in putrescine, and a stimulation of several neurotransmitter systems, including the catecholamines, DA and NE. The last indicates an interaction between the two branches of the stress response.

Few data exist as yet on the effects of CRF, but it is already clear that this peptide is capable of eliciting a powerful sympathetic activation, in addition to triggering an HPA response.

Function of the neurochemical stress responses

An important question is the extent to which the neurochemical changes that have been observed during stress may underlie the adaptation of the organism to stress. The increases in biosynthetic enzymes for the catecholamines permit coping with increased demands on the noradrenergic neurons, but do not in themselves permit adaptation. The down-regulation of adrenergic receptors that occurs in chronic stress presumably represents an adaptation of the brain to deal with excess NE release, and this could explain the behavioral adaptation. On the basis of similarities in the effects of chronic stress and tricyclic antidepressant therapy on adrenergic receptors, Stone has postulated that similar mechanisms may be involved. A simple-minded view would suggest that prophylactic antidepressant therapy could protect against the effects of stress, and indeed there is some evidence to support this idea. However, another view suggests that the situation is far more complex. Weiss has postulated that behavioral (motor) deficits in stressed animals (so-called "learned helplessness") are associated with deficits in noradrenergic function (correlated with depletions of cerebral NE, especially in locus coeruleus).

A full understanding of the stress response awaits knowledge of the true biological role of catecholaminergic systems. Although NE has been postulated to play a role in neuroplasticity and has long been implicated in higher nervous functions such as learning and memory, it also appears to play relatively mundane roles in limiting blood flow and regulating the blood-brain barrier.

Further reading

Axelrod J, Reisine TD (1984): Stress hormones: Their interaction and regulation. *Science* 224:452–459

Dunn AJ (1984): Effects of ACTH, β-lipotropin, and related peptides on the central nervous system. In: *Peptides, Hormones and Behavior*, Nemeroff CB, Dunn AJ, eds. New York: Spectrum Publications, pp 273–348

Dunn AJ, Kramarcy NR (1984): Neurochemical responses in stress: Relationships between the hypothalamic-pituitary-adrenal and catecholamine systems. In: *Handbook of Psychopharmacology, Vol 18*, Iversen LL, Iversen SD, Snyder SH, eds. New York: Plenum Press, pp 455–515

Mason JW (1971): A re-evaluation of the concept of "non-specificity" in stress theory. *J Psychiat Res* 8:323–333

Stone EA (1975): Stress and catecholamines. In: *Catecholamines and Behavior, Vol 2*, Friedhoff AJ, ed. New York: Plenum Press, pp 31–72

Substance Abuse

Steven M. Mirin

For centuries individuals have ingested, smoked, inhaled, or injected chemical substances that alter thinking and mood. For the vast majority of these individuals experimental or recreational drug use is a transient phenomenon. For a small minority, however, use becomes, over time, abuse, with harmful effects on physical, emotional, social, and occupational functioning.

In general, drugs subject to abuse exert direct pharmacologic effects on the brain that result in the alteration of mood. Moreover, there is substantial evidence that the reinforcing properties of these drugs can also be demonstrated in animals. For example, acute administration of drugs like heroin or cocaine will produce a brief, but memorable, euphoria which motivates the user to repeat the experience. Monkeys exposed to injections of these drugs will perform operant work such as bar pressing in order to obtain more injections, even in the absence of physical dependence or drug withdrawal symptoms.

In addition to the recreational use of drugs for their primary reinforcing effects, psychoactive drugs are also used by some individuals to self-medicate feelings of anxiety or depression. Thus, a chronically anxious individual may find that the antianxiety effects of a benzodiazepine or the sedative effects of alcohol may provide effective, but temporary, relief. Similarly a depressed individual who is experiencing feelings of helplessness and hopelessness may find that the self-administration of a central nervous system (CNS) stimulant like cocaine will temporarily reverse this negative feeling state.

Regardless of the initial motivation for use, repetitive self-administration of psychoactive drugs can have serious adverse consequences. Among these are the development of psychological dependence, drug tolerance, and physical dependence. Psychological dependence refers to a felt need to use a particular drug (or class of drugs) because its pharmacologic effects are deemed essential for normal psychological functioning. If denied access to their drug of choice, psychologically dependent individuals experience considerable anxiety, fearing the loss of what they consider to be an important tool for coping with internal or external stress. As a result, they will take steps to assure access to their drug even if it means forgoing their usual social or work obligations.

With some drugs of abuse such as opiates and CNS depressants and stimulants repetitive use is accompanied by the development of tolerance, a condition in which increasingly larger doses of a particular drug must be taken in order to reproduce the drug's original pharmacologic effects. Drug tolerance may result from an increased rate of drug metabolism, or a change in the sensitivity of certain target cells to the drug's pharmacologic effects. In addition, individuals who develop tolerance to one drug often manifest tolerance to other drugs in the same pharmacologic class. For example, chronic users of CNS depressants like alcohol may manifest tolerance to other CNS depressants such as the benzodiazepines and hypnotic-sedatives. In general, the rate of tolerance development is a function of the drug consumed, the doses used, the frequency of administration, and the response the drug elicits in those cells responsible for its metabolism and those cells through which its pharmacologic effects are mediated.

Drug tolerance is frequently, but not invariably, accompanied by physical dependence. This implies a state of altered cellular physiology developed in response to repetitive drug use. Consequently abrupt, or even gradual, withdrawal of a dependence-producing drug may result in the development of a characteristic abstinence syndrome. Dramatic abstinence syndromes may be seen following abrupt withdrawal of opiates and CNS depressants and, to a lesser extent, CNS stimulants and marijuana. The signs and symptoms of drug withdrawal are usually the antithesis of those pharmacologic effects originally produced by the drug. For example, heroin addicts who experience tension relief and euphoria and who manifest pinpoint pupils and decreased gastrointestinal motility during episodes of drug administration respond to heroin withdrawal with anxiety, depression, dilated pupils, and diarrhea.

With this as an introduction, the following sections will briefly review each of the major classes of commonly abused drugs with emphasis on the relationship between their effects on the brain and their abuse potential.

The opiates

Pharmacology. The opiates include the natural alkaloids of opium such as morphine and codeine, the semisynthetic derivatives such as heroin and hydromorphine, and purely synthetic agents such as meperidine and methadone. Opiate drugs may be ingested or administered intravenously or intramuscularly. Intravenous administration produces a rapid elevation in the blood levels of these drugs, and the blood level achieved appears to be correlated with the degree of intoxication produced. As with most psychoactive drugs the opiates appear to exert their effects on the brain and other organ systems by binding to receptor sites on cell membranes. These so-called opiate receptors, which have been identified in various areas of the brain as well as in the gastrointestinal tract, may also be the binding sites for naturally occurring opioid peptides called enkephalins and endorphins.

Acute opiate intoxication is characterized by a brief but intense euphoria, followed by relaxation, drowsiness, mental clouding, apathy, and motor retardation. Somatic effects include slowing of the gastrointestinal tract, respiratory depression, and a drop in pulse and blood pressure. In cases of intentional or unintentional overdose, death may occur as a result of respiratory depression or cardiovascular collapse.

Repetitive use of opioid drugs leads to the development of

tolerance to their euphorigenic, analgesic, sedative, and respiratory depressant effects and, in some cases, profound physical dependence. Conversely, abrupt withdrawal of these drugs in physically dependent individuals leads to the development of a characteristic abstinence syndrome whose symptoms essentially reflect increased CNS and autonomic nervous system arousal. The onset, duration, and severity of this abstinence syndrome depends on the particular opiate used, the degree of tolerance and physical dependence developed, the time elapsed since the last dose, and the psychological response of the patient to the physical discomfort of drug withdrawal. Thus, for a relatively short-acting drug like heroin, withdrawal symptoms may begin as soon as 6 hours after the last dose and peak 48 to 72 hours later. In contrast, withdrawal from methadone, a longer acting agent, usually begins 24 to 48 hours after the last dose and may persist over a 3- to 5-week period. Withdrawal symptoms can be alleviated by renewed administration of any opiate, taking advantage of the cross tolerance that exists between these drugs. In treatment settings, methadone, a synthetic opioid with a relatively long duration of action (i.e., 24 to 36 hours) is helpful in this regard. Indeed, the binding of methadone to opiate receptor sites not only relieves withdrawal symptoms but will also block the euphoria which accompanies acute administration of drugs like heroin. Chronic use of methadone in this fashion is accompanied by the development of tolerance to the sedative, analgesic, and depressant effects seen during the acute phase of treatment. Consequently, chronic maintenance on high doses of methadone is an effective and popular treatment for recidivist opiate addicts who would otherwise relapse to the use of illicit heroin or other shorter acting opioids.

Another drug used to ameliorate the symptoms of opiate withdrawal is clonidine, a nonopiate antihypertensive agent which acts by stimulating alpha-2 receptors located in the neurons of the locus ceruleus (LC) region of the midbrain. This region gives rise to long neuronal processes that form the noradrenergic pathways of the CNS. Clonidine-induced stimulation of alpha-2 noradrenergic receptors reduces the firing rate of LC neurons through a type of feedback inhibition, thus decreasing noradrenergic activity in both the CNS and the periphery. Since many of the symptoms of opiate withdrawal are due to noradrenergic hyperactivity, clonidine is an effective treatment for opiate withdrawal symptoms.

Use and abuse of opiate drugs. In addition to the development of tolerance and physical dependence, the chronic user of opiate drugs often experiences the psychosocial sequelae of drug dependence. The latter may include disruption of interpersonal ties, occupational failure, and the development of a deviant lifestyle necessary to support the illicit use of these drugs. Treatment approaches are customarily aimed at controlling drug use and helping individuals to develop an alternative lifestyle. Pharmacologic approaches to limiting illicit drug use include the use of methadone and narcotic antagonist drugs, which are structurally similar to the opiates and bind tightly to opiate receptor sites, blocking the pharmacologic effects of subsequently administered opiate drugs. In individuals who are physically dependent, administration of narcotic antagonists such as naltrexone or naloxone will precipitate a severe withdrawal syndrome.

Among the nonpharmacologic approaches to the treatment of chronic opiate users are individual, group, and family therapy. In patients with a history of frequent relapse, prolonged residence in a drug-free therapeutic community may help the individual focus on learning to cope with life's travails without the use of mind-altering chemicals.

The CNS depressants

Pharmacology. The CNS depressants include the barbiturates, sedative-hypnotics, and benzodiazepines. These drugs share the common ability to induce sedation or sleep by depressing CNS activity. In addition, the benzodiazepines appear to have specific antianxiety effects. Though the precise mechanism of action of the CNS depressants is not well worked out, some drugs in this group, particularly the benzodiazepines, probably alter the activity of those CNS neurons which use gamma-aminobutyric acid (GABA) as a neurotransmitter.

CNS depressants may be administered orally or parenterally and are well absorbed from the gastrointestinal tract. Their onset of action is dependent, in part, on their lipid solubility, which affects their ability to cross cell membranes. Their duration of action is determined by the rate at which they are distributed to various body tissues as well as the rate of metabolic degradation and renal excretion.

Intoxication with CNS depressants is a direct result of the effects of these agents on the cerebral cortex and brain stem. Severe intoxication is usually the result of purposeful overdose, often in a suicide attempt. Overdose cases are complicated by the fact that simultaneous administration of more than one CNS depressant may overwhelm available liver enzymes, resulting in a clinical effect which is greater than the sum of the effect of either agent administered alone. The self-administration of such drug combinations (e.g., alcohol with a benzodiazepine) is an important cause of overdose deaths.

With repetitive use of CNS depressants, both tolerance and physical dependence develops. Tolerance appears to be of two types, metabolic tolerance, in which liver enzymes are elaborated, allowing for more rapid metabolism of these drugs, and pharmacodynamic tolerance in which the body tissues appear to develop a subsensitivity to their effects. Finally, tolerance development to a particular CNS depressant (including alcohol) confers some degree of cross tolerance to other drugs in this general class.

Use and abuse of CNS depressants. Federal statistics indicate that more than 100 million prescriptions are written each year for CNS depressant drugs. Of these, the vast majority are for the benzodiazepines like diazepam, chlordiazepoxide, and flurazepam, but there are a host of other drugs with sedative and antianxiety effects that are in wide use. In general, these drugs are initially prescribed for the symptomatic treatment of insomnia and anxiety. Unfortunately, prolonged administration of these agents leads to the rapid development of tolerance to both their hypnotic and anxiolytic effects. Consequently, there is often a tendency to gradually increase the dose over time with the subsequent development of even greater tolerance and physical dependence.

Some individuals abuse depressant drugs such as barbiturates or sedative-hypnotics intermittently during sprees of intoxication. Others abuse depressants in an attempt to alleviate the symptoms of withdrawal from other drugs such as opiates or alcohol. Abusers of such CNS stimulants as amphetamine and cocaine may also use alcohol, sedative-hypnotics, or benzodiazepines in an attempt to modify stimulant-induced jitteryness, hyperactivity, and insomnia.

Most patients who have been taking CNS depressants in moderate to high doses over more than several weeks will experience at least mild abstinence symptoms if the drugs are withdrawn abruptly. Abstinence symptoms may include a rebound increase in rapid eye movement (REM) sleep with increased dreaming, middle-of-the-night awakening, irritability, and anxiety. In more severe cases, hypotension, hyperthermia, delirium, seizures, and death, due to cardiovascular collapse, may occur. The onset of withdrawal symptoms is, in part, dependent on the duration of action of the drug used. Thus abrupt withdrawal of a short-acting agent like pentobarbital may be followed by abstinence symptoms within 12 to 16 hours. In contrast, withdrawal of a longer acting drug, like diazepam, may not produce withdrawal symptoms until 7 to 10 days have elapsed. Severe depressant withdrawal syndrome should be treated in a hospital setting. The primary goal of treatment is to reduce CNS irritability. This is usually done by reinstituting treatment with a CNS depressant and gradually withdrawing the agent under medical supervision.

Treatment of depressant abuse should attempt to address root causes. In this regard, some depressant abusers suffer from underlying psychiatric disorders including depression, panic and anxiety states, and a host of characterologic problems. Thus, treatment should be directed at the alleviation of these problems through the use of individual, group, and family therapy and, where necessary, psychotropic medication.

The CNS stimulants

Pharmacology. CNS stimulants like amphetamine, methamphetamine, and cocaine exert their stimulatory effects on the brain by facilitating the release of the neurotransmitter, norepinephrine (NE) from CNS neurons. They also increase the availability of this neurotransmitter at functionally important receptor sites in the CNS by interfering with the reuptake, and subsequent metabolic breakdown, of NE by CNS neurons. The stimulant drugs also inhibit monoamine oxidase, an enzyme important in the degradation of NE. In high doses, these drugs also appear to exert similar effects on dopamine-containing neurons. This may contribute to their tendency to produce psychosis in chronic, high-dose users.

In general, administration of CNS stimulants produces increased mental alertness, wakefulness, a sense of initiative and confidence, and elevated mood. The drugs also suppress both appetite and REM sleep. Acute stimulant intoxication is characterized by restlessness, hyperactivity, irritability, talkativeness, anxiety, and labile mood. At higher doses, these drugs induce agitation, tremor, confusion, and dysphoria. In addition, some individuals become suspicious and even paranoid. In general, the intoxication syndrome is self-limited and wanes as the drugs are metabolized. With its extremely short duration of action (20 minutes or less), cocaine inspires a tendency toward repeated use in order to maintain the drug-induced euphoria.

With repetitive use, tolerance and mild physical dependence develop to most of the effects of CNS stimulants. Under these conditions abrupt discontinuation of drug use may be followed by a characteristic withdrawal syndrome marked by lethargy, fatigue, and depression. These withdrawal effects, in turn, may precipitate further drug use.

Use and abuse of CNS stimulants. Though stimulants like amphetamine and methylphenidate have some role in the short-term treatment of obesity and the longer term treatment of narcolepsy and attention deficit disorder, they also have significant potential for abuse. Thus, some individuals take these drugs chronically for their mood-elevating effects while others engage in sporadic, high-dose use in search of intoxication and euphoria. This last motivation is particularly common among those who self-administer methamphetamine and cocaine.

In the 1960s, amphetamine and methamphetamine were popular drugs of abuse. More recently, however, these drugs have been replaced in popularity by cocaine, the abuse of which is becoming a major public health problem. Cocaine is most commonly self-administered intranasally (snorted) although intravenous use and "freebasing" (i.e., smoking of the basified extract) are increasingly common as well. The vast majority of cocaine users do so for social or recreational purposes. Among chronic high-dose users, however, recent studies have demonstrated that there is a substantial minority who are attempting to self-treat an underlying psychiatric disorder; most commonly depression or manic depressive illness.

Finally, chronic high-dose use of amphetamine, methamphetamine, or cocaine may be accompanied by the development of a toxic psychosis characterized by suspiciousness, fear, increased aggression, delusions of persecution, ideas of reference, and visual and auditory hallucinations. Superficially this state resembles paranoid schizophrenia. Withdrawal of the drug of abuse is usually followed by recovery, though some patients continue to manifest psychotic symptoms for months afterward. Chronic stimulant abuse usually requires treatment in an inpatient setting. Rebound depression may be treated with psychotherapeutic support and in some cases with antidepressant drugs. Avoidance of relapse requires that the underlying psychiatric and characterologic disorders found in these patients be addressed as well.

The hallucinogens

Pharmacology. A wide variety of agents derived from either natural sources (primarily plants) or synthesized in the laboratory have the ability to induce perceptual distortions and hallucinations in those who either ingest them or administer them intravenously. These drugs have been variously labeled psychedelic (i.e., mind expanding) or psychotomimetic (i.e., mimicking psychosis); however, neither term precisely describes their effects.

The commonly abused hallucinogens fall into two chemical classes: The indolealkylamines include LSD, psilocybin, and dimethyltryptamine (DMT) and are structurally similar to the neurotransmitter 5-hydroxytryptamine (serotonin). The phenylethylamines include mescaline and dimethoxy methylamphetamine (DOM) and are structurally related to the neurotransmitters norepinephrine and dopamine.

Use and abuse of hallucinogens. Since the hallucinogens have no well-defined therapeutic use, all use may be defined as abuse. In humans the hallucinogens induce a toxic psychosis characterized by perceptual changes (e.g., illusions and hallucinations), confusion between sensory modalities (synesthesias), a subjective sense that time is slowed, loss of body and ego boundaries, and emotional lability. The duration of hallucinogen intoxication varies from 1 to 24 hours depending on the agent used. Aftereffects, like psychic numbness, may last for days, however.

For some hallucinogen users drug-induced perceptual changes, coupled with a generalized loss of body and ego boundaries, is extremely anxiety provoking. Consequently, the most common adverse reaction following the use of LSD or other hallucinogens is acute panic, during which the user fears that he is "losing his mind." Panic symptoms may be alleviated by the presence of supportive friends and a decrease in the amount of sensory stimulation. Another adverse reaction to hallucinogen use is the development of a toxic delirium characterized by hallucinations, delusions, agitation, disorientation, and paranoia. In individuals with underlying mood disorders or schizophrenia, hallucinogen use may exacerbate the illness. Finally, some users have recurrent drug-like experiences (i.e., flashbacks) that may occur for weeks to months following even the one-time use of a hallucinogenic substance.

Though not much is known about the psychological sequelae of chronic hallucinogen use, it is the general clinical impression that such individuals are unusually passive and suffer from a variety of personality disorders.

Phencyclidine (PCP)

Pharmacology. Phencyclidine, also known as PCP or "angel dust," is an animal tranquilizer which has achieved a fair degree of popularity as a drug of abuse. The drug has stimulant, depressant, hallucinogenic and analgesic effects. It can be taken orally, intranasally, and intravenously, and it can be inhaled through the lungs when smoked. It is metabolized by the liver and stored in fatty tissues. At low doses, the serum half-life of PCP is approximately 45 minutes. However, following administration of large doses, the half-life may be as long as 3 days due to sequestration in fatty tissue.

In low doses, PCP is a CNS depressant. It produces impaired perception, incoordination, generalized numbness, and a blank-stare appearance. At higher doses, there is more disruption of CNS functioning, with slurred speech, ataxia, increased deep tendon reflexes, and catalepsy. Large doses may also produce hypertension, seizures, respiratory depression, coma, and death.

Use and abuse of PCP. The abuse of PCP apparently stems from the drug's ability to induce feelings of disassociation and perceptual distortions. Acute intoxication, which may last 4 to 6 hours after a single dose, is often accompanied by increased sensitivity to external stimuli and elevated mood. Some patients exhibit bizarre posturing or catatonia as well.

PCP use may also produce a psychotic state that can persist for days or weeks. Patients with a prior history of psychosis appear to be particularly vulnerable to this complication. The psychosis itself is characterized by increased motor activity, insomnia, agitation, paranoid delusions, and increased sensitivity to external stimuli. Assaultive behavior is not uncommon. In the absence of specialized medical treatment, the psychotic episode may resolve over a period of 5 to 15 days.

Finally, there is some evidence that chronic PCP use can induce profound anxiety, impairment of cognitive functioning, social isolation and withdrawal, and unpredictable violence.

Marijuana

Pharmacology. Of all the drugs mentioned here, with the exception of alcohol, marijuana is probably the most widely used. Surveys suggest that upward of 60 million people over age 13 have tried the drug at least once. The major active ingredient in marijuana is delta-9-tetrahydrocannabinol (THC). There is approximately 5–20 mg of THC in an average marijuana cigarette, about half of which is absorbed into the bloodstream during the smoking process. Oral administration of marijuana results in somewhat lower absorption of THC, as well as slower onset of action.

Once absorbed, THC is stored in fatty tissues and some is bound to plasma protein. Metabolism is via liver enzymes, followed by excretion through the gastrointestinal tract and kidneys. As a result of sequestration in fatty tissue, some THC can be found in plasma for as long as 6 days after the administration of a single dose.

The effects of marijuana intoxication are colored by the dose consumed, its route of administration, and other physiological and interpersonal variables. The expectations of the user and the setting in which the drug is consumed are extremely important in this regard. The most common effects are feelings of well-being, euphoria, and relaxation, along with altered time sense and a heightened awareness of one's environment. Thought processes are slowed and short-term memory is impaired. Some users feel that they derive special insights while intoxicated and even trivial events may have profound meaning in the intoxicated state. Daily use of marijuana results in the development of mild tolerance and physical dependence. In chronic, heavy users, abrupt withdrawal produces restlessness, irritability, disturbed sleep, anorexia, sweating, tremor, and gastrointestinal upset.

Use and abuse of marijuana. Though the vast majority of individuals who try marijuana may be classified as experimenters, approximately 35% become regular users and about 5% become daily users of the drug. Among the latter group, one finds an increase in the prevalence of both drug-induced and nondrug psychopathology when compared to experimenters and casual users. Moreover, chronic heavy users also tend to abuse other psychoactive drugs, including alcohol.

In inexperienced users, acute intoxication may precipitate a panic state which may simulate acute psychosis, though reality testing remains intact. As in the case of the hallucinogens, large doses may also produce toxic delirium marked by confusion, feelings of derealization, paranoia, visual and auditory hallucinations, and dysphoric mood. In general, both panic and toxic delirium are time-related, lasting for a few hours to a few days, and waning as the drug is gradually metabolized.

Chronic heavy marijuana use may be accompanied by what some authors have described as an "amotivational syndrome" characterized by decreased drive, diminished attention span, poor judgment, apathy, and impaired communication skills. Though they may continue to perform tasks that do not require the ability to concentrate, social, occupational, and interpersonal functioning are impaired. Chronic high dose marijuana use may also lead to the development of a psychotic state which clinically looks very much like PCP psychosis. This condition has been described primarily in heavy users residing in countries where hashish, a more potent, concentrated form of marijuana, is used.

Summary

The propensity for any individual to abuse psychoactive drugs is the result of developmental, social, cultural, and interpersonal factors. Personal attitudes toward drug taking and drug intoxication, the attitudes of one's peer group toward such behavior, and the availability of the drug itself are also important. Moreover, in some instances genetic and biological vulnerability also plays a role. The ability of these drugs to induce euphoria and relieve tension is extremely reinforcing in both animals and humans and sometimes leads to repetitive use with the concurrent development of tolerance and physical dependence. Moreover, once individuals become physically dependent, continued drug taking becomes necessary in order to avoid the development of abstinence symptoms. Finally, there is also evidence that abstinence symptoms may be experienced by previously dependent individuals as the result of behavioral conditioning. All these factors make substance abuse an extremely complicated behavior to change. Consequently, successful treatment of such individuals requires the development of a multimodal, eclectic approach which takes this complexity into account.

Further reading

Bourne PG (1976): *Acute Drug Abuse Emergencies*. New York: Academic Press

Grinspoon L, Bakalar JB (1976): *Cocaine. A Drug and Its Social Evolution*. New York: Basic Books

Meyer RE, Mirin SM (1979): *The Heroin Stimulus*. New York: Plenum Press

Mirin SM, Meyer RE (1978): The Treatment of Substance Abusers. In: *Principles of Psychopharmacology*, Clark WG, del Guidice J, eds. New York: Academic Press

Tolerance and Physical Dependence

William R. Martin

Tolerance and physical dependence are complex biological responses to the chronic administration of drugs and are probably intimately related to basic neurophysiological processes such as habituation and accomodation. Chemical stimuli and receptive mechanisms were probably among the first sensory mechanisms to evolve to assist in satisfying food needs and in avoiding toxic chemicals. To facilitate acquisitive and avoidance movements, negative feedback mechanisms probably developed at a very early stage of evolution. Further, a large portion, perhaps all, of the communication between neurons in the central nervous system is conducted using chemical (neurotransmitters, modulators, and hormones) stimuli and receptors. Finally, virtually all drugs that induce tolerance and dependence exert their actions, either directly or indirectly, by altering neurotransmitter-receptor interactions (Table 1).

Chronic administration of drugs diminishes some of their major effects. For example, 30 mg morphine may produce severe respiratory depression in some subjects; however, in individuals who have become tolerant to morphine, a dose of 60 mg will not overtly or markedly alter respiration. However, the respiratory function of morphine-tolerant subjects is not normal; their rate is somewhat slowed and the ability of CO_2 to stimulate respiration is diminished. Similarly the ability of morphine to produce feelings of well-being are diminished by repeated administration, and subjects who receive large doses of morphine chronically become hypophoric. Animal studies have explored the effects of chronic administration of morphine on its ability to diminish responses to painful stimuli and have shown that the dose-response curves in naive and tolerant animals are parallel but the curve of the tolerant animal is shifted to the right. Evidence from both humans and animals indicates that both cognitive processes and learning are involved in the development of drug tolerance. Thus many individuals who are particularly susceptible to the depressant effects of alcohol may avoid complex motor function and hence partially conceal their inebriation.

Tolerance to some actions of drugs may develop rapidly and can be demonstrated after a single dose. Because of the rapidity with which it develops it is called acute tolerance. Tolerance to other actions of drugs may require many large doses administered for several days. Because of these characteristics it is called chronic tolerance.

Because of these diverse characteristics of tolerance to the actions of drugs, it would seem probable that tolerance is not one phenomenon but several and that there are probably several mechanisms involved in its development. Many species decrease the effect of drugs by metabolizing them more rapidly with a liver microsomal enzyme system comprised of many closely related enzymes with different specificities. Some drugs metabolized by these enzyme systems also induce or increase the amount of the enzyme, thus enhancing the inactivation of the drug and related drugs and reducing their effects. This is called dispositional or metabolic tolerance and probably accounts for about half the tolerance induced by chronic administration of pentobarbital. Tolerance that cannot be accounted for by increased metabolism of drugs is called functional tolerance, and for most drugs that alter central nervous system function, this is thought to be mediated by changes in brain function.

The degree of tolerance to some effects of central nervous system drugs is ambiguous since the effects of drugs in tolerant animals is different than in naive animals. For those effects where tolerance can be measured unambiguously the degree of tolerance induced varies greatly among drugs. The degree of tolerance can be estimated by calculating a tolerance index—the ratio of the amount of drug necessary to produce an effect of a certain magnitude in tolerant and nontolerant animals or tissue. Tolerance indices of between 10 and 100 and even higher have been observed for some LSD-like hallucinogens and for some opioids, and the tolerance index for pentobarbital is between 2 and 3. These observations also suggest that the brain has several mechanisms for developing tolerance. These mechanisms have provided a tool for classifying drugs with similar mechanisms of action using the phenomena of cross tolerance. Cross tolerance is said to exist when the induction of tolerance to one drug confers tolerance to another drug. This technique has been used extensively to show that several chemicals of diverse structure share a common mode of action with LSD. Similar studies have been conducted using subtypes of opioids.

Physical dependence is a group of phenomena characterized by the emergence of a unique pattern or syndrome of signs and symptoms when the drug is withdrawn from subjects who have ingested several or many doses or is displaced from its receptors by a competitive antagonist. When the drug is with-

Table 1. Types of Drugs of Abuse, the Neurotransmitters that they Mimic or Modulate, and the Receptors with Which they Interact

Drug Type	Neurotransmitter	Receptor and Receptor Subtypes
Opioid analgesics	endorphins, enkephalins, and dynorphins	μ, κ, δ, ϵ, σ
Alcohol		GABA
Tobacco	acetylcholine	nicotinic cholinergic and other nicotinic receptors
Amphetamines		dopaminergic and noradrenergic
Benzodiazepines		GABA
Sedative-hypnotics		GABA
LSD-like hallucinogens	serotonin and tryptamine	serotonin and tryptamine

drawn the resulting abstinence syndrome is called withdrawal abstinence; when an antagonist is used to precipitate abstinence the resulting syndrome is called precipitated abstinence. The withdrawal and precipitated abstinence syndrome usually are similar but may differ in certain respects.

Tolerance and physical dependence are thought to be particularly pernicious aspects of certain drug actions for three reasons. (1) The degree of tolerance and the level of dependence are thought to be related to the amount of drug administered and the frequency and duration of administration. Thus when tolerance begins to develop, the amount of drug required to produce the same therapeutic effect is larger, which in turn increases both tolerance and physical dependence. (2) Symptoms of abstinence are uncomfortable to the patient. When the patient learns that the symptoms of abstinence can be relieved by the dependence-producing drug, the drug-procuring activity "hustling" is established, which may lead to antisocial behavior. (3) The continuing use of drugs of abuse may produce long-lasting changes in brain function called protracted abstinence, which may also contribute to relapse of detoxified addicts. Protracted abstinence impacts negatively with personality diatheses that predispose individuals to use drugs that produce feelings of well-being.

Various types of drugs produce different types of physical dependence. Tolerance and physical dependence produced by opioids have been most carefully studied. It is known that a clinically significant degree of physical dependence in humans can be produced by small doses of morphine (30 mg/day) administered for about a month. Further, an abstinence syndrome can be precipitated after only a single large dose or several smaller doses of methadone. Thus physical dependence can be induced rapidly and with only small doses of morphine-like drugs. Opioid receptors are distributed throughout the brain and opiate drugs alter the function of most regions of the brain. Thus it is not surprising that signs of abstinence involve many physiological systems. Further signs of opioid abstinence are opposite to those of opioid primary effects: (1) Thus morphine depresses spinal cord reflexes evoked by painful stimuli and pain perception; these reflexes are overly responsive to painful stimuli in abstinence, and patients complain of joint and muscle pain. (2) Morphine depresses the medullary respiratory center's responsivity and sensitivity to carbon dioxide; its responsivity and sensitivity is enhanced during abstinence. (3) Morphine constricts pupils by stimulating the mesencephalic Edinger-Westfal nucleus; pupils are dilated during abstinence. (4) Morphine depresses the release of luteinizing hormone by acting on the hypothalamus and this is associated with a decrease in libido; abstinence is associated with high levels of plasma luteinizing hormone and testosterone and spontaneous ejaculations occur. The morphine abstinence syndrome involves enhancement of spinal cord nociceptive reflexes; medullary hyperexcitability is indicated by tachycardia, tachypnea, increased blood pressure, and in part by increased arousal and nausea and vomiting; at a mesencephalic level by mydriasis; and at a hypothalamic level by altered hormonal function and hyperthermia. There are many changes in affect produced by opiates. In nondependent opiate abusers affect is characterized by feelings of well-being, enhanced self-image, and feelings of effectiveness in interpersonal interactions. When patients are abstinent from opiates they feel weak, tired, apathetic, withdrawn, unpopular, and ineffective. They overreact to the discomfort associated with painful sensations, nausea, vomiting, and intestinal cramps. Many of these feelings are probably mediated through higher centers. Thus a schizophrenic patient who was withdrawn from morphine following a prefrontal lobotomy did not complain even though he exhibited typical signs of opiate withdrawal.

When patients are chronically administered depressant drugs such as alcohol, barbiturates, and antianxiety drugs, some degree of tolerance develops to their effects; however, as with dependence on opioids, the patient ingesting depressants is far from normal. They may show signs of drunkenness and incoordination, thinking may be slowed, and mood may be labile, ranging from feelings of well-being to hostility, aggressiveness, irritability, and depression. Chronic use of depressants may markedly inhibit rapid eye movement sleep. Less is known about signs and symptoms of abstinence in patients who are dependent on alcohol, barbiturates, and other sedative-hypnotics and antianxiety drugs. Among well-identified signs and symptoms are tremor, weakness, involuntary twitching, nausea, vomiting, loss of appetite, grand mal seizures, orthostatic hypotension, dizziness, insomnia, anxiety, and delirium that may or may not be associated with fever. Rapid eye movement sleep may be markedly increased. The abstinence syndrome associated with dependence on depressants is much more life-threatening than the abstinence syndrome associated with opioid dependence. The ability of various depressants to substitute for each other in dependent patients has not been well investigated. Drugs such as pentobarbital and paraldehyde will clearly suppress the abstinence syndrome in alcohol and barbiturate dependence; however, in some patients very large doses may be required to suppress seizures and delirium. Not all signs and symptoms emerge at the same time. Tremors almost invariably appear during the first 24 hours of abstinence. Epileptic phenomena usually do not appear until the second day, and delirium usually does not occur until the fourth to seventh day of abstinence. It is not known why different signs and symptoms of abstinence from depressants such as alcohol and barbiturates appear at different times. Although the rate of elimination of the drug may play a role, other processes may be important. The discovery of competitive antagonists of the benzodiazepine depressant has shed some light on this issue. When antagonists are administered to benzodiazepine-dependent animals to precipitate abstinence, tremor appears within 10 to 15 minutes; however, several hours may elapse before seizure activity becomes manifest. This time course of benzodiazepine abstinence is quite distinct from abstinence precipitated by opioid antagonists in opioid-dependent subjects. The opioid abstinence syndrome appears in its complete and maximally intense form within a few minutes after the administration of the antagonists and starts declining in intensity shortly thereafter. These observations suggest that opioid and depressant dependence differ in their basic mechanisms.

Not all drugs that induce central nervous system functional tolerance produce the same degree of physical dependence. Thus the chronic ingestion of large doses of marihuana or hashish induces tolerance but only a mild degree of physical dependence characterized by nausea, vomiting, diarrhea, loss of appetite, tremor, sweating, irritability, restlessness, and sleep disturbances. On the other hand, LSD-like hallucinogens that produce profound changes in perceptions and moods as well as marked changes in autonomic and motor reflex function do not induce physical dependence.

Tolerance and physical dependence have captured the imagination of many investigators who have attempted to understand these two phenomena. Insofar as is known, opioid drugs such as morphine depress nerve function by acting as agonists which decrease neuronal function and excitability by increasing K^+ and/or decreasing calcium flux across certain nerve membranes. Depressants such as barbiturates and benzodiazepines are thought also to decrease neuronal excitability by increasing the effects of the inhibitory nerve transmitter gamma-aminobutyric acid (GABA) which increases Cl^- flux across nerve cell membranes. It is important to recognize that the two groups of drugs that induce tolerance and the most overt types of physical dependence enhance endogenous inhibitory transmitter activity. Further, the two types of inhibitory processes that are mimicked or enhanced—opioidergic and GABA-ergic—are different in several ways in that they involve different processes, are involved in different neuronal circuits, and hence produce different physiological changes. Further, both types of inhibitory processes are ubiquitous in the nervous system.

It is therefore not surprising that dependence on these two types of drugs creates abstinence syndromes with many signs and symptoms. Drugs that produce excitatory effects such as the LSD-like hallucinogens and amphetamines, although capable of inducing tolerance, do not produce physical dependence characterized by a severe abstinence syndrome. This suggests that the clinically significant types of physical dependence are a consequence of the brain's effort to adapt to central inhibitory processes. The nature of these adaptive processes is still uncertain; however, all theories of physical dependence postulate negative feedback mechanisms of one sort or another that operate on receptors, neurotransmitters, other pathways, or homeostats. To state this general hypothesis in specific terms for different levels of organization the following formulations have been proposed. If nerve cell receptors are not stimulated by endogenous or exogenous agonists resulting in a decrease in the formation of active drug-receptor complexes, several up-regulatory processes may be recruited. The resulting decrease in active drug receptor complexes may be a consequence of the destruction of presynaptic elements either by lesions or chemicals, drug or neurotransmitter-induced depression of presynaptic element function, or the occupation of postsynaptic receptors by antagonist. Thus the adaptive mechanisms recruited to counteract the depression induced by increased activity of inhibitory processes must occur at the inhibited neuron or secondary neurons whose activity is decreased. It is important to recognize that inhibitory agonists such as opioids and depressants that act directly on receptors or modulate receptors must continue to exert their inhibitory activity in the tolerant and dependent animal since their chronic administration continues to suppress the abstinence syndrome.

Figure 1 illustrates some of the postulated mechanisms whereby inhibitory drugs can evoke tolerance and dependence. Assume that drug d inhibits neuron A which would result in decreased transmitter release at synapse A-B. This effect would have the following actions: (1) There is up-regulation of the receptors responsive to the neurotransmitter whose release has been depressed (neurons B and D). (2) The amount of negative feedback from neuron B to A is decreased as a consequence of the decreased excitation of B. This results in a relative increase in activity of neuron A and an increase in transmitter synthesis by neuron A. Both effects would result in tolerance; the increased neurotransmitter synthesis, in physical dependence. Redundant pathways (C-D) may also mediate the same function and be under negative feedback control from neuron B. Drug d would then increase the activity of neuron C, resulting in increased synthesis of the neurotransmitters by neuron C and give rise to both tolerance and physical dependence. When drug d is not present or withdrawn, the activity of neuron D would be markedly above its activity prior to the administration of drug d (abstinence with physical dependence) because of receptor up-regulation and neurons B and D and transmitter up-regulation at neurons A and C.

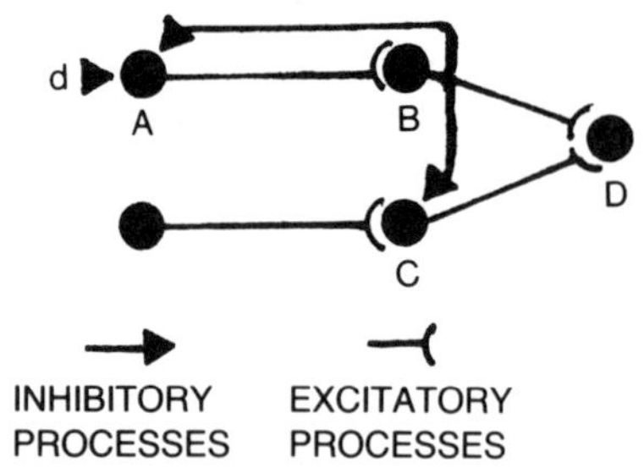

Figure 1. Schematic diagram of a neuronal net illustrating the effect of an inhibitory or a depressant drug in up-regulating receptors and neurotransmitters, resulting in the production of tolerance and dependence.

Further reading

Jaffe JH (1980): Drug addiction and drug abuse. In: *The Pharmacological Basis of Therapeutics,* 6th ed, Gilman AG, Goodman LS, Gilman A, eds. New York: Macmillan

Jones RT (1980): Human effects: An overview. NIDA Research Monograph 31, 54–80

Kalant H, LeBlanc AE, Gibbins RJ (1971): Tolerance to, and dependence on, some non-opiate psychotropic drugs. *Pharmacol Rev* 23(3):135–191

Martin WR (1982): Hypnotics. In: *Psychotropic Agents Part III,* Hoffmeister F, Stille G, eds. Berlin: Springer-Verlag

Martin WR, Sloan JW (1977): Neuropharmacology and neurochemistry of subjective effects, analgesia, tolerance, and dependence produced by narcotic analgesics. In: *Drug Addiction I,* Martin WR, ed. Berlin: Springer-Verlag

Wikler A (1968): *The Addictive States.* Baltimore: Williams & Wilkins

Contributors

David M. Asher Laboratory of Central Nervous System Studies, National Institute of Neurological and Communicative Disorders and Stroke, National Institutes of Health, Bethesda, Maryland 20205, U.S.A.
Creutzfeldt-Jakob Disease

M. Avoli Department of Neurology and Neurosurgery, McGill University, Montreal Neurological Institute, Montreal, Quebec, H3A 2B4, Canada
Epilepsy

Stephanie J. Bird Science, Technology, and Society Program and The Center for Technology, Policy and Industrial Development, Massachusetts Institute of Technology, Cambridge, Massachusetts 02139, U.S.A.
Premenstrual Syndrome

Lyle W. Bivens Director, Division of Basic Sciences, National Institute of Mental Health, Rockville, Maryland 20857, U.S.A.
Psychosurgery

Charles Brenner 1040 Park Avenue, New York, New York 10028, U.S.A.
Psychoanalysis

Jack R. Cooper Department of Pharmacology, Yale University School of Medicine, New Haven, Connecticut 06510, U.S.A.
Neuropharmacology

Joseph T. Coyle Division of Child Psychiatry, The Johns Hopkins Medical Institutions, Baltimore, Maryland 21205, U.S.A.
Alzheimer's Disease

John W. Crayton Department of Psychiatry, The University of Chicago, Chicago, Illinois 60637, U.S.A.
Mental Illness, Nutrition and

Adrian J. Dunn Department of Neuroscience, JHM Health Center, University of Florida College of Medicine, Gainesville, Florida 32610, U.S.A.
Stress, Neurochemistry of

Terrance M. Egan University Laboratory of Physiology, Parks Road, Oxford, England
Morphine

Charles J. Epstein Departments of Pediatrics and of Biochemistry and Biophysics, University of California, San Francisco, California 94143-0106, U.S.A.
Down Syndrome

Max Fink Department of Psychiatry, School of Medicine, SUNY at Stony Brook, Long Island, New York 11794, U.S.A.
Convulsive Therapy

Arnold J. Friedhoff Department of Psychiatry, New York University School of Medicine, Millhauser Laboratories, 550 First Avenue, New York, New York 10026, U.S.A.
Gilles de la Tourette Syndrome

Albert M. Galaburda Neurological Unit, K-4, Beth Israel Hospital, Boston, Massachusetts 02215, U.S.A.
Dyslexia

Alan J. Gelenberg Psychiatrist-in-Chief, The Arbour, Jamaica Plain, Massachusetts 02130, U.S.A. and Chief, Special Studies Clinic, Massachusetts General Hospital, Boston, Massachusetts 02115, U.S.A., and Associate Professor of Psychiatry, Harvard Medical School, Boston, Massachusetts 02115, U.S.A.
Mood Disorders

Clarence J. Gibbs, Jr. Laboratory of Central Nervous System Studies, National Institute of Neurological and Communicative Disorders and Stroke, National Institutes of Health, Bethesda, Maryland 20205, U.S.A.
Creutzfeldt-Jakob Disease

P. Gloor Professor, Department of Neurology and Neurosurgery, McGill University, Montreal-Neurological Institute, Montreal, Quebec, H2A 2B4, Canada
Epilepsy

Donald W. Goodwin Department of Psychiatry, University of Kansas Medical Center, Kansas City, Kansas 66103, U.S.A.
Alcoholism

John H. Greist Department of Psychiatry, University of Wisconsin, Center for Health Sciences, Madison, Wisconsin 53792, U.S.A.
Lithium in Psychiatric Therapy

James F. Gusella Director, Neurogenetics Laboratory, Massachusetts General Hospital, Boston, Massachusetts 02114, U.S.A.
Huntington's Disease (HD)

Joseph E. Hawkins, Jr. Professor Emeritus, Otorhinolaryngology; Kresge Hearing Research Institute, University of Michigan Medical School, Ann Arbor, Michigan 48109, U.S.A.
Deafness

Philip S. Holzman Department of Psychology, Harvard University, Cambridge, Massachusetts 02138, U.S.A.
Eye Movement Dysfunctions and Mental Illness

C.B. Ireland Professor of Psychiatry, Department of Psychiatry, and Neurosciences Program, University of Alabama at Birmingham, Birmingham, Alabama 35294, U.S.A.
Hallucinogenic Drugs

Susan D. Iversen Neuroscience Research Centre, Merck Sharp & Dohme Research Laboratories, Terlings Park, Harlow, Essex CM20 2QR, England
Psychopharmacology

James W. Jefferson Department of Psychiatry, University of Wisconsin, Center for Health Sciences, Madison, Wisconsin 53792, U.S.A.
Lithium in Psychiatric Therapy

Harold Kalant Professor, Department of Pharmacology, University of Toronto, Toronto, M5S 1A8, Canada and Director, Biobehavioral Research Department, Addiction Research Foundation, Toronto, M5S 2S1, Canada
Addiction

Conan Kornetsky Boston University School of Medicine, Boston, Massachusetts 02118, U.S.A.
Heroin (Diacetylmorphine)

Harriet O. Kotsoris Department of Neurology, The New York Hospital–Cornell Medical Center, New York, New York 10021, U.S.A.
Coma

William R. Martin Professor and Chairman, Department of Pharmacology, University of Kentucky College of Medicine, Lexington, Kentucky 40536, U.S.A.
Tolerance and Physical Dependence

Steven Matthysse The Mailman Research Center, McLean Hospital, Belmont, Massachusetts 02178, U.S.A.
Mental Illness, Genetics of

Neal E. Miller Professor Emeritus, Rockefeller University, 1230 York Avenue, New York, New York 10021, and Research Affiliate, Yale University, New Haven, Connecticut 06520, U.S.A.
Behavioral Medicine

Steven M. Mirin Associate Clinical Professor of Psychiatry, Harvard Medical School, Medical Director, Westwood Lodge Hospital, Westwood, Massachusetts 02090, U.S.A.
Substance Abuse

Hugo W. Moser Director, John F. Kennedy Institute for Handicapped Children and Professor of Neurology and Pediatrics, The Johns Hopkins University, Baltimore, Maryland 21205, U.S.A.
Mental Retardation

Dennis L. Murphy Chief, Laboratory of Clinical Science, NIMH, Bethesda, Maryland 20895, U.S.A.
Monoamine Oxidase (MAO) Inhibitors in Psychiatric Therapy

William L. Nyhan Department of Pediatrics, School of Medicine, University of California, San Diego, La Jolla, California 92093, U.S.A.
Phenylalanine and Mental Retardation (PKU)

Edward M. Ornitz Department of Psychiatry, Division of Mental Retardation and Child Psychiatry, and Brain Research Institute, University of California, Los Angeles, California 90024, U.S.A.
Autism

Herbert Pardes Professor and Chairman, Department of Psychiatry, Columbia University, College of Physicians and Surgeons, New York, New York 10032, U.S.A.
Psychiatry, Biological

Robert C. Petersen Formerly Assistant Director of Research, National Institute on Drug Abuse, Rockville, Maryland 20957, U.S.A.
Marijuana
Phencyclidine

Fred Plum Chairman, Department of Neurology, Cornell University Medical College, New York, New York 10021, U.S.A.
Coma
Dementia

Gardner C. Quarton University of Michigan, Ann Arbor, Michigan 48107, U.S.A.
Psychosis

Domeena C. Renshaw Professor, Department of Psychiatry, Director, Sexual Dysfunction Clinic, Loyola University of Chicago, Maywood, Illinois 60153, U.S.A.
Eating Disorders

Leo J. Reyna Behavioral Sciences Center, Nova University, Fort Lauderdale, Florida 33314, U.S.A.
Behavior Therapy, Applied Behavior Analysis, and Behavior Modification

Elliott Richelson Professor and Consultant, Departments of Psychiatry and Psychology and of Pharmacology, Mayo Clinic and Foundation, Rochester, Minnesota 55905, U.S.A
Antidepressants

Timothy Roehrs Sleep Disorders and Research Center, Henry Ford Hospital, Detroit, Michigan 48202, U.S.A.
Sleep Disorders

Thomas Roth Sleep Disorders and Research Center, Henry Ford Hospital, Detroit, Michigan 48202, U.S.A.
Sleep Disorders

Arnold B. Scheibel Departments of Anatomy and Psychiatry and Brain Research Institute, UCLA Medical Center, Los Angeles, California 90024, U.S.A.
Aging of the Brain

David V. Sheehan Professor of Psychiatry and Director of Clinical Research, University of South Florida College of Medicine, Tampa, Florida 33612, U.S.A.
Anxiety and Anxiety Disorders

Kathy H. Sheehan Department of Psychiatry, University of South Florida College of Medicine, Tampa, Florida 33612, U.S.A.
Anxiety and Anxiety Disorders

William J. Shoemaker Director, Neurobiology Laboratory, Department of Psychiatry, University of Connecticut Health Center, Farmington, Connecticut 06032, U.S.A.
Fetal Alcohol Syndrome

John R. Smythies Department of Psychiatry, and Neurosciences Program, University of Alabama at Birmingham, Birmingham, Alabama 35294, U.S.A.
Hallucinogenic Drugs

Solomon H. Snyder Departments of Neuroscience, Pharmacology and Experimental Therapeutics, Psychiatry and Behavioral Sciences, The Johns Hopkins University School of Medicine, Baltimore, Maryland 21205, U.S.A.
Neuroleptic Drugs

Larry R. Squire Veterans Administration Medical Center, San Diego, California 92161, U.S.A., and Department of Psychiatry, University of California School of Medicine, La Jolla, California 92093, U.S.A.
Amnesia

Erik Strömgren Professor of Psychiatry, Institute of Psychiatric Demography, Psychiatric Hospital, DK-8240 Risskov, Denmark
Schizophrenia

Norman J. Uretsky Division of Pharmacology, College of Pharmacy, Ohio State University, Columbus, Ohio 43210, U.S.A.
Amphetamines

Roger D. Weiss Assistant Professor of Psychiatry, Harvard Medical School, Boston, Massachusetts 02115, U.S.A. and Director, Alcohol and Drug Abuse Treatment Center, McLean Hospital, Belmont, Massachusetts 02178, U.S.A.
Cocaine

Frank Zorick Sleep Disorders and Research Center, Henry Ford Hospital, Detroit, Michigan 48202, U.S.A.
Sleep Disorders